THE CONCEPT OF MULTIVERSE

The concept of a multiverse is one that has captured the human imagination for centuries. It's an idea that challenges our understanding of reality and forces us to question everything we thought we knew about our existence. But what exactly is a multiverse? And how does it fit into our current understanding of the universe?

In simple terms, the multiverse theory suggests that there isn't just one universe (the one we inhabit), but an infinite number of universes existing side by side, each with its own laws and properties. These universes are often referred to as "parallel" or "alternate" universes.

Imagine if you will, standing on the edge of a vast ocean. Each wave in this ocean represents a different universe within the multiverse. Some waves are larger than others, some crash against each other while others flow harmoniously side by side – these variations represent the differences between each individual universe.

But where did this concept originate from? The idea can be traced back to ancient philosophy and mythology - from Hindu cosmology's endless cycles of creation and destruction to Norse legends' nine realms co-existing simultaneously.

However, it wasn't until quantum mechanics came onto the scene in the early 20th century that scientists began seriously considering such possibilities. Quantum mechanics introduced concepts like superposition (where particles can exist in multiple states at once) and entanglement (where particles become interconnected regardless of distance), which seemed to suggest that reality was far more complex than previously imagined.

Our main character in this journey through parallel worlds is not unlike you or me - curious about his place in this grand scheme called life; eager yet apprehensive about venturing into unknown territories; seeking answers yet humbled by new questions arising at every turn.

As he embarks on his quest across dimensions, he'll encounter strange phenomena beyond comprehension - time flowing backwards; gravity acting repulsively instead attractively; worlds where dinosaurs never went extinct or where the Axis powers won World War II. He'll meet alternate versions of himself - some leading drastically different lives due to choices made differently at critical junctures.

Through his eyes, we'll explore philosophical questions that arise from multiverse theory: If there are infinite universes, is there a version of me who's achieved everything I've ever dreamed of? Does fate exist or do we have free will? What does death mean in an infinite multiverse?

As you journey through this book, remember that it's not just about understanding scientific theories but also about reflecting on our own existence and purpose. It's about embracing uncertainty and finding comfort in the vastness of it all.

To end this chapter, here's a thought experiment for you: Imagine waking up tomorrow in a parallel universe. Everything seems exactly the same except for one minor detail – your favorite color has changed. How would this affect your perception of reality? Would it make you question your sanity or accept the possibility of other realities existing alongside ours?

In the next chapter, we'll delve into history tracing back how humans first conceived such mind-boggling ideas as multiple universes co-existing simultaneously.

HISTORICAL OVERVIEW OF THE MULTIVERSE THEORY

The concept of a multiverse, an infinite ensemble of universes parallel to our own, is not new. It has been part and parcel of human imagination for centuries, finding its roots in ancient philosophy and mythology before evolving into a serious scientific proposition.

In the beginning was Parmenides, an ancient Greek philosopher who lived around 500 BC. He proposed that reality is unchanging and uniform; there are no differences or changes in space or time. This idea can be seen as one of the earliest precursors to the modern concept of a multiverse.

Fast forward to medieval times when philosophers like Thomas Aquinas grappled with questions about God's omnipotence – could he create other worlds? The debate continued through Renaissance thinkers such as Giordano Bruno who suggested that stars were distant suns surrounded by their own planets - implying multiple realities.

However, it wasn't until the advent of quantum mechanics in the early 20th century that we began to see concrete scientific theories supporting this notion. In his doctoral thesis at Princeton University in 1957, Hugh Everett III proposed what would later become known as "the many-worlds interpretation" (MWI) of quantum physics. According to MWI every possible outcome for each event defines or exists in its universe.

Everett's theory initially met with skepticism from mainstream physicists including Niels Bohr and Werner Heisenberg whose Copenhagen Interpretation dominated quantum mechanics at that time. However, over years it gained traction among some scientists due to its ability to explain certain phenomena without resorting to wave function collapse - a controversial process which Copenhagen Interpretation couldn't satisfactorily define.

Parallelly during mid-1980s cosmologist Andrei Linde developed "eternal chaotic inflation" theory suggesting our universe is just one bubble within an infinite cosmic foam where new universes constantly form driven by laws potentially different from ours.

The 21st century saw the rise of string theory, which posits that our universe exists on a brane (short for membrane) within higher-dimensional space. This gave birth to the concept of "brane worlds", where multiple universes exist parallel to each other in different dimensions.

Today, multiverse theories continue to evolve and diversify. Max Tegmark's four levels of multiverses propose an increasingly complex hierarchy from Level I (regions beyond our cosmic horizon with similar physical laws) up to Level IV (different mathematical structures govern these universes).

Our protagonist, embarking on this journey through parallel realities, stands on the shoulders of these intellectual giants. As he navigates through infinite possibilities and mind-bending concepts, he will grapple not only with scientific complexities but also philosophical quandaries about fate, free will and his own existence.

As we delve deeper into this exploration let's remember physicist Richard Feynman's words: "I think nature's imagination is so much greater than man's; she's never going to let us relax." The adventure has just begun!

Exercise: Reflect upon your understanding of reality before reading this chapter. How does it compare or contrast with historical perspectives? What questions do you have now about the nature of reality?

FROM FICTION TO SCIENCE - POP CULTURE'S INFLUENCE ON MULTIVERSE THEORY

In the realm of science fiction, parallel universes have long been a staple. They provide an endless playground for imagination and creativity, allowing writers to explore alternate histories, mirror worlds where evil twins reside or even realities where toast always lands butter-side up. But what if these seemingly fantastical ideas weren't confined to the pages of comic books or flickering onto our screens in late-night sci-fi marathons? What if they were grounded in scientific theory?

The concept of multiple universes, or a 'multiverse', has gradually moved from the fringes of physics into its mainstream over recent decades. This shift can be attributed not only to advancements in theoretical physics but also significantly influenced by pop culture's fascination with the idea.

Take for instance H.G Wells' "The Door in The Wall", published back in 1906; it toyed with the idea of parallel dimensions accessible through portals hidden within our world. Or consider Philip K Dick's novel "The Man In The High Castle" which explores an alternate reality where World War II was won by Axis powers.

These works and many others like them have played a crucial role in popularizing multiverse concepts among masses and making such abstract theories more palatable. They've sparked curiosity about life beyond our universe and inspired generations of scientists who grew up reading these stories.

One such example is Dr Hugh Everett III who proposed his Many-Worlds Interpretation (MWI) back in 1957 after being deeply fascinated by science fiction literature during his formative years. His groundbreaking work suggested that every quantum event creates new branches leading to different outcomes – essentially creating infinite parallel universes!

Pop culture continued this symbiotic relationship with science when

Star Trek introduced audiences worldwide to various forms of multiverses throughout its series run since 1966—be it Mirror Universes, Parallel Dimensions or Temporal Anomalies. This not only entertained millions but also subtly educated them about complex scientific theories.

The influence of pop culture on multiverse theory isn't just historical; it's ongoing. The Marvel Cinematic Universe, with its recent foray into the multiverse concept in films like "Doctor Strange" and series like "Loki", has brought the idea to an even wider audience, making conversations about quantum physics part of everyday discourse.

But why is this important? Because science doesn't exist in a vacuum. It's influenced by society and culture, just as much as it influences them. Pop culture acts as a bridge between common people and scientists, translating abstract concepts into relatable narratives that inspire curiosity and drive exploration.

As we delve deeper into understanding our universe (or should we say 'multiverse'), let us remember that every great scientific discovery began as a spark of imagination - often kindled by stories told around campfires or written down in books or shown on screens big and small.

So next time you watch your favorite sci-fi show or read a comic book featuring parallel universes, remember: you're not just consuming entertainment; you're participating in the age-old human tradition of storytelling that drives our collective quest for knowledge.

To conclude this chapter here are some questions to ponder upon:

1) How does popular media shape your perception of complex scientific theories?

2) Can you think of any other examples where fiction has inspired real-life scientific advancements?

3) What role do you believe pop-culture plays in advancing our understanding of theoretical concepts?

* * *

In the next chapter, we'll dive deeper into understanding dimensions and parallel universes from a more technical perspective – so buckle up!

UNDERSTANDING DIMENSIONS AND PARALLEL UNIVERSES

In the grand tapestry of existence, we are but a single thread. We exist in three dimensions: length, width, and height. These dimensions form the framework of our reality—a stage where life's drama unfolds. But what if there were more to this stage than meets the eye? What if beyond our perception lie other realms—parallel universes existing in different dimensions?

To understand parallel universes, we must first grasp the concept of dimensions. A dimension is essentially a measurable extent such as length or width that defines an object's position within a given space.

The first dimension is simply a line—a straight path with no depth or height—only length. Add another perpendicular line to it; you get two-dimensional space like a flat sheet of paper with length and breadth but no depth—an entity living here would only perceive lines and shapes without any sense of solidity.

Now add depth by drawing another line at right angles to these two lines—you've just created three-dimensional space--the world as humans perceive it—with objects having length, breadth, and height.

But what about the fourth dimension? This is where things start getting interesting—and tricky! The fourth dimension isn't spatial—it's temporal—time itself! Imagine being able to move not just forward-backward-left-right-up-down—but also past-future!

This brings us to parallel universes—or multiverses—as they're often called. If time can be considered as an additional axis along which one could travel (just like moving left or right), then each point on this axis represents an entirely different universe—one that exists parallel to ours!

Our hero for this journey into understanding multiverses will be Sam —a curious young scientist who stumbles upon these concepts while researching quantum physics.

Sam had always been fascinated by science fiction stories involving alternate realities and multiple timelines—he never thought he'd find himself exploring similar ideas in his research. But as he delved deeper into quantum mechanics, he found that the line between science fiction and reality was blurrier than he'd ever imagined.

Sam's journey began with a simple question: "What if?" What if there were more dimensions beyond our perception? What if these dimensions housed other universes—each with its unique laws of physics, histories, and life forms?

As Sam pondered over these questions, his research took him on an intellectual adventure through the realms of string theory and M-theory—proposals in theoretical physics suggesting the existence of multiple dimensions (up to 11 according to some theories!).

These theories opened up a whole new world for Sam—a world where our universe was just one among countless others floating like bubbles in a cosmic ocean—an idea known as the multiverse hypothesis.

But understanding this concept wasn't easy—it challenged everything Sam knew about reality. He had to let go of his preconceived notions and open his mind to possibilities that seemed straight out of a sci-fi novel!

To help you embark on your own journey towards understanding parallel universes, here are some exercises:

1) Visualize Dimensions: Start by visualizing lower dimensions (a point for zero dimension; a line for one dimension; square for two

dimensions). Then try imagining adding another perpendicular direction each time—you'll find it gets trickier as you move higher!

2) Read Up On Theories: Dive into string theory or M-theory—these can be complex but don't worry about understanding every detail—the goal is to get familiar with concepts like extra-dimensions.

3) Explore Sci-Fi Literature/Films: Many works explore ideas related to parallel universes—they can serve as great thought experiments helping visualize such abstract concepts.

Remember—as we delve further into this topic—we're not merely exploring scientific theories—we're venturing into uncharted territories of human imagination! So buckle up—it's going to be a wild ride!

QUANTUM MECHANICS AND ITS ROLE IN THE MULTIVERSE

In our journey through the multiverse, we must first understand one of the most fundamental theories that underpin it - quantum mechanics. This chapter will delve into this fascinating world where particles can be in two places at once, cats can be both dead and alive (Schrodinger's cat), and observing a particle changes its state.

Quantum mechanics is a branch of physics that deals with phenomena on a very small scale, such as molecules, atoms, and subatomic particles like electrons and photons. It was developed in the early 20th century by scientists who were trying to explain behaviors that couldn't be accounted for by classical physics.

One key concept in quantum mechanics is superposition. Superposition refers to an object being able to exist simultaneously in multiple states until observed or measured when it collapses into one state. For example, an electron orbiting an atom doesn't have a definite position until you measure it; instead, there's only a probability distribution of where you might find it.

* * *

Now imagine applying this principle not just to tiny particles but also to entire universes! This leads us directly into what physicist Hugh Everett III proposed as the Many-Worlds Interpretation (MWI) of quantum mechanics back in 1957. According to MWI every time there's more than one possible outcome – from atomic level events up to human decisions – all outcomes occur across different branches of reality creating parallel universes.

Our hero Sam Doe had always been fascinated by science fiction stories about parallel worlds since he was young boy reading comic books late at night under his blanket with flashlight on hand. He never thought those wild ideas could actually hold some truth until he stumbled upon Everett's theory during his college years studying Physics.

Sam found himself captivated by these concepts which seemed so outlandish yet scientifically plausible according him "It felt like I was living inside my favorite sci-fi novel, except this was real. It was mind-blowing."

To illustrate the concept of superposition and MWI, let's consider a simple example: flipping a coin. In our everyday experience, we know that when you flip a coin it will either land heads or tails. But in the quantum world, until you actually observe the result of the flip, the coin is both heads and tails at once - it's in a state of superposition.

When applied to MWI theory, each possible outcome corresponds to different universes. So when you flip that quantum coin there's one universe where it lands on heads and another parallel universe where it landed on tails.

Now imagine every decision you've ever made as these flips of quantum coins creating countless versions of reality with different versions of yourself living out all possible outcomes simultaneously across multiple universes!

* * *

As Sam delved deeper into his studies he began questioning everything around him "If I chose chocolate ice cream instead vanilla last night did another version me enjoy vanilla somewhere else? What about major life decisions like choosing my college major or moving cities?"

This chapter has introduced some complex concepts but don't worry if they seem overwhelming at first! Quantum mechanics is famously quoted by Richard Feynman as something nobody truly understands due its counter-intuitive nature.

Exercise for readers: Reflect upon your own life decisions big small how would things be different if had chosen differently? Imagine what those alternate realities might look like are they drastically different from current reality or perhaps not so much?

In next chapters we'll continue exploring more about multiverse theories including cosmic inflation string theory dark matter energy stay tuned journey just beginning!

THE HERO'S JOURNEY BEGINS – AN UNEXPECTED DISCOVERY

Our story begins in the most ordinary of places, a small town where nothing ever seems to change. Our hero, let's call him Sam, is an average man leading an average life. He works as a librarian and spends his free time reading books on theoretical physics and cosmology - subjects that have fascinated him since childhood.

One day while sorting through old books in the library basement, he stumbles upon a dusty tome titled "The Multiverse: A Guide for Travelers". Intrigued by its title and unusual appearance, he decides to take it home for further examination.

As Sam delves into the book at home later that night, he finds himself engrossed in tales of parallel universes and alternate realities. The book talks about quantum mechanics and string theory but presents them not as abstract scientific concepts but as practical tools for traversing across different dimensions.

Suddenly something strange happens; the words start glowing on one of the pages. As if guided by some unseen force, Sam reads aloud an incantation written there. Suddenly everything around him starts spinning wildly before coming back into focus again...but something has changed drastically!

Sam finds himself standing not in his cozy living room but what appears to be another world entirely! It looks similar yet different from our own Earth with subtle differences like unfamiliar constellations twinkling brightly above or peculiar looking plants growing around.

This unexpected discovery marks the beginning of Sam's journey across multiple universes - each unique with its laws of physics and reality constructs.

But how does one navigate such uncharted territories? How can you survive when even basic rules we take granted may no longer apply?

Well dear reader this is exactly what we are going to explore together along with our protagonist! We will learn about various theories proposed by scientists over centuries which might help us understand these new realms better.

For instance consider Schrödinger's cat experiment. It suggests that until observed, particles can exist in multiple states simultaneously. Now imagine if this concept applied to entire universes!

Or take the theory of cosmic inflation which proposes that our universe is just one bubble among countless others floating in a vast cosmic ocean.

But remember these are not mere academic exercises for Sam anymore but tools for survival and exploration!

As we journey along with him through different dimensions, we will encounter strange worlds and even stranger beings. We might come across civilizations far advanced than ours or primitive ones where time seems to have stood still.

We will face challenges testing our courage, intellect and resilience; from solving complex quantum puzzles to battling inter-dimensional creatures.

And who knows maybe somewhere along the way we might also find answers to some of our deepest existential questions like why do we exist? Is there a purpose behind all this?

So buckle up dear reader as we embark on an adventure unlike any other! Remember though - once you start seeing reality as fluid rather than fixed, there's no going back...

The hero's journey has begun – an unexpected discovery leading us into uncharted territories filled with wonder, danger and infinite possibilities...

DIVING INTO INFINITY - EXPLORING INFINITE UNIVERSES

The concept of infinity is a mind-boggling one. It's like trying to imagine the end of the universe, or what existed before time began. The human brain struggles with such vastness and eternity because it operates within finite parameters. But when we talk about multiverses, we're talking about infinite possibilities.

Imagine standing on the edge of a cliff overlooking an ocean that stretches as far as your eyes can see in every direction. This is how our protagonist felt at this point in his journey – he was about to dive into infinity.

* * *

Our hero had always been fascinated by the mysteries of life and existence. He spent countless hours pondering questions like "Why are we here?" "What else exists beyond our known universe?" His thirst for knowledge led him down various paths, from studying quantum physics to exploring ancient philosophies and religions.

One day while conducting an experiment involving quantum entanglement (a phenomenon where particles become interconnected regardless of distance), something extraordinary happened – he stumbled upon a portal leading to another universe!

This wasn't just any other universe; it was strikingly similar yet subtly different from ours—a parallel reality where history took a slightly different course—like viewing an alternate version of your favorite movie with different actors playing familiar roles.

Now let's pause for a moment here and consider this scenario: What if you could visit an infinite number of universes? Each one unique, each one presenting new opportunities for exploration and discovery?

In these infinite universes, there would be worlds where dinosaurs never went extinct, worlds ruled by advanced AI systems or civilizations far more technologically advanced than ours. There might even be universes where magic exists!

But diving into infinity isn't without its challenges—it requires courage, curiosity, resilience...and perhaps most importantly—an open mind ready to embrace whatever comes along.

As our hero prepared himself mentally for this journey, he couldn't help but feel a mix of excitement and trepidation. He was about to embark on an adventure that would take him beyond the boundaries of his known reality.

He knew there would be risks involved—after all, exploring unknown territories always comes with potential dangers. But he also understood that without taking risks, one cannot make significant discoveries or advancements.

So how does one navigate through infinite universes? This is where our understanding of quantum mechanics comes into play. According to the many-worlds interpretation (MWI) proposed by physicist Hugh Everett III in 1957, every possible outcome of each event defines or exists in its own "world" or universe.

In other words, for every decision you make—no matter how insignificant it may seem—a new universe branches out from your current one where you made a different choice. It's like standing at a crossroads with an infinite number of paths stretching out before you!

Now imagine if you could travel between these parallel realities at will! That's precisely what our hero set out to do.

As we follow his journey through these countless worlds, we'll encounter strange creatures and civilizations; witness spectacular cosmic events; explore exotic landscapes—and perhaps even come face-to-face with alternate versions of ourselves!

But remember: while this might sound like pure science fiction now — it wasn't too long ago when ideas like space travel and artificial intelligence were considered far-fetched fantasies as well...

The exploration of multiverses presents us with endless possibilities for discovery and learning—but it also raises profound questions about existence itself: What does it mean to be human? How do we define reality? And ultimately—is there any limit to what we can achieve?

Our protagonist's dive into infinity serves as both a metaphorical journey towards self-discovery and an actual physical voyage across multiple dimensions—an epic quest that promises thrilling adventures ahead...

To conclude this chapter let me leave you with some food for thought:

If you had the opportunity to explore infinite universes, where would you go? What kind of worlds would you like to discover?

In the next chapter, we'll delve deeper into the concept of cosmic inflation and bubble universes. But for now, let's take a moment to marvel at the sheer vastness and complexity of our multiverse—and dream about the endless adventures that await us in this grand cosmic playground!

COSMIC INFLATION AND BUBBLE UNIVERSES

In the grand tapestry of cosmic theories, one stands out for its audacity, its creativity, and its sheer mind-boggling implications. This is the theory of cosmic inflation – a concept that has revolutionized our understanding of how the universe came to be.

Imagine you're standing on an empty stage in a vast auditorium. Suddenly, within less than a blink of an eye, this stage expands exponentially until it's larger than the observable universe. That's what cosmic inflation proposes happened at the birth of our universe - an incredibly rapid expansion from something infinitesimally small to something astronomically large.

But where do bubble universes fit into all this? Let's take a journey through time and space to find out.

Our hero had just begun his exploration into parallel worlds when he stumbled upon these concepts. He was initially overwhelmed by their complexity but soon realized they were key pieces in understanding his new reality.

Cosmic inflation suggests that during the first tiny fraction of a second after Big Bang, space expanded faster than light speed due to quantum fluctuations. The energy driving this process was hypothesized as 'inflation' or 'inflaton field'. When this field decayed (like radioactive material), it gave rise to matter and radiation leading to Big Bang's heat - marking end of inflation era.

* * *

However, if we consider quantum mechanics principles here too; there could have been regions where inflaton didn't completely decay - causing eternal or ongoing inflation creating multiple "bubble" universes with different physical properties!

This idea forms basis for Andrei Linde's chaotic inflation model suggesting multiverse as Swiss cheese-like structure with each hole representing separate bubble universe including ours!

Now imagine our hero finding himself in such another bubble universe where gravity repels instead attracts! Or perhaps one where time runs backward! These possibilities might sound like science fiction but are potential realities according to multiverse theory.

But how can we test these theories? One way is through cosmic microwave background (CMB) radiation - the afterglow of Big Bang. Some scientists believe that if our universe has collided with another bubble universe in past, it would have left detectable patterns in CMB!

Our hero was fascinated by this idea and decided to look for such evidence in his interdimensional travels. He knew it wouldn't be easy but he was ready for the challenge.

Now, let's take a moment here. Imagine yourself as an explorer like our hero. What kind of bubble universes would you want to discover? How different do you think they could be from ours?

As a thought experiment, try visualizing your own version of a parallel universe using principles discussed so far. Write down its properties and compare them with those of our known universe.

Remember, when dealing with concepts as vast and complex as cosmic inflation and bubble universes; it's okay not fully understand or even feel overwhelmed at times! Even greatest minds grapple with these ideas!

* * *

In words of physicist Richard Feynman: "If you think you understand quantum mechanics, then you don't." So keep questioning, keep exploring because that's what science –and indeed any adventure into unknown– is all about!

Next time when we meet again on this journey across multiverses; we'll delve deeper into string theory & brane worlds adding more layers to our understanding! Until then remember: The cosmos no longer ends where eyesight fails; instead it extends beyond horizons of imagination!

STRING THEORY AND BRANE WORLDS

In the grand tapestry of the cosmos, there are threads that weave together to form a picture far more complex than we could ever imagine. These threads, according to string theory, are not particles or waves but tiny strings of energy vibrating at different frequencies. This chapter will delve into the fascinating world of string theory and its implications for our understanding of parallel universes.

The concept behind string theory is both simple and mind-bogglingly complex. At its core, it suggests that everything in our universe - from galaxies to atoms - is composed of these infinitesimal strings. The various forms matter takes on are simply due to different modes of vibration in these strings.

But where do branes come into play? In string theory, 'branes' (short for membranes) represent multidimensional objects within which these strings exist and vibrate. Our entire universe might be just one such 3-dimensional 'brane', floating amidst an array of other branes in higher dimensional space – a concept known as M-theory.

To illustrate this idea further, let's take an example from pop culture: consider the Upside Down realm from Stranger Things series; it exists simultaneously with our reality but on another plane entirely – much like how another universe could exist on a separate brane.

* * *

Now imagine our protagonist John Doe stumbling upon this revelation during his journey through multiverses. He finds himself standing before two mirrors facing each other creating infinite reflections - a metaphorical representation for endless parallel universes existing side by side yet separated by their respective dimensions or 'branes'.

This discovery would dramatically change John's perception about his place in the cosmos; he was no longer confined within one universe but had access to countless others! It would also present him with new challenges as navigating between these brane worlds would require understanding laws governing higher dimensional spaces.

Let's pause here for some reflection: How does the idea of our universe being just one among countless others on different branes make you feel? Does it excite you with possibilities or overwhelm you with its scale?

Now, let's move onto some practical exercises to help consolidate this concept. Imagine yourself in John's shoes and write a short story about your first encounter with another brane world. What would it look like? How would the laws of physics operate there?

In conclusion, string theory and the concept of brane worlds open up new dimensions (literally!) in our understanding of multiverses. They challenge us to rethink our notions about reality and inspire us to dream beyond what we currently know.

As physicist Brian Greene once said, "String theory elegantly offers a way to unify the two great but seemingly contradictory pillars of 20th-century physics - quantum mechanics and general relativity." And who knows, perhaps one day we might be able to prove its predictions about parallel universes too!

THE MANY-WORLDS INTERPRETATION OF QUANTUM

MECHANICS

Our journey into the multiverse has taken us through dimensions, parallel universes, and even quantum mechanics. Now we delve deeper into one of the most fascinating interpretations in quantum physics - the Many-Worlds Interpretation (MWI).

The MWI was first proposed by physicist Hugh Everett III in 1957. It suggests that all possible alternate histories and futures are real, each representing an actual "world" or "universe." In other words, it postulates a vast number of universes where every possible outcome to every event exists.

Imagine flipping a coin. According to classical physics, there are two outcomes – heads or tails. But according to MWI, when you flip that coin, reality splits into two distinct worlds: one where the coin lands on heads and another where it lands on tails.

Now let's bring back our hero from previous chapters who is navigating his way through these multiple realities. He flips a cosmic coin in Universe A; it comes up heads. However simultaneously he also finds himself in Universe B where the same coin shows tails! This might sound like science fiction but remember what Arthur C Clarke said? Any sufficiently advanced technology is indistinguishable from magic!

This concept can be mind-boggling because it challenges our everyday experience of reality as singular and linear. Yet this interpretation solves many paradoxes inherent in traditional quantum mechanics such as Schrödinger's infamous cat being both dead and alive until observed.

To illustrate further consider this scenario: Our hero encounters an alien civilization with technology far beyond ours capable of manipulating probabilities at will using their understanding of MWI! They play games with causality itself making impossible things happen just for fun!

* * *

But how does this affect our hero's journey? Well imagine if he could tap into these infinite possibilities too! What if he could choose which world to inhabit based on desired outcomes? He could become a king, a pauper, a scientist or an artist in different worlds. The possibilities are endless!

Now let's take this concept and apply it to our own lives. What if every decision we make creates alternate realities where different versions of ourselves exist? This thought can be empowering but also overwhelming.

So here's an exercise for you: Think about a major decision you made recently. Now imagine what your life would look like had you chosen differently. That reality might just exist in another universe according to MWI!

As we delve deeper into the multiverse theory, remember that these concepts are not meant to confuse or frighten us but rather expand our understanding of reality itself. As physicist Brian Greene said "Exploring the unknown requires tolerating uncertainty."

In the next chapter, we will follow our hero as he encounters his first parallel universe and navigates through different realities using his new found knowledge of quantum mechanics and MWI.

Remember - science is not only about finding answers; it's also about asking better questions! So keep questioning as we journey together through multiple universes!

THE HERO'S FIRST ENCOUNTER WITH A PARALLEL UNIVERSE

Our hero Sam had always been an explorer at heart. His curiosity was insatiable and his thirst for knowledge unquenchable. But nothing could have prepared him for the journey he was about to embark on.

* * *

Sam had spent years studying quantum mechanics and string theory, engrossed in the idea of parallel universes. He'd read every book available on the subject, attended countless lectures, and even dedicated his doctoral thesis to exploring the theoretical existence of multiverses. Yet all this time it remained just that - a theory.

One day while working late in his lab at MIT, something extraordinary happened. A series of unexpected results from an experiment led Sam to stumble upon what appeared to be a portal into another universe.

As he stepped through this shimmering gateway between realities, he found himself standing in an alternate version of Boston – familiar yet strikingly different. Buildings were taller and more futuristic; people moved around using personal flying devices; there were no cars but rather advanced magnetic levitation trains zipping across beautifully designed tracks suspended high above ground level.

This wasn't just another city or country; it was clearly another world entirely – one where technological advancements far surpassed those of our own reality.

The sight was overwhelming but exhilarating! Imagine being Christopher Columbus discovering America or Neil Armstrong setting foot on the moon for the first time!

But how did this happen? How did science fiction become reality?

Let's rewind back to Sam's lab where everything started...

Sam had been experimenting with superstring vibrations trying to prove their role in creating multiple dimensions as suggested by M-theory when suddenly there was a power surge causing unusual fluctuations within his equipment which seemed to tear open space-time itself creating what looked like a wormhole!

Now you might ask "Is such thing even possible?" Well according to physicist Kip Thorne if enough energy is concentrated in a small enough region, it could theoretically create a wormhole. And that's exactly what happened!

But let's get back to Sam...

As he navigated through this parallel universe, he noticed not just technological differences but also social and cultural ones. There was an air of harmony and peace; people seemed happier, healthier, more content.

He saw schools where children were taught meditation alongside math; hospitals where doctors used advanced AI for precise diagnoses and treatments; parks filled with lush greenery amidst skyscrapers indicating a perfect balance between urbanization and nature preservation.

It was as if this world had taken all the best aspects of our society while eliminating its flaws. It gave him hope seeing how humanity could evolve given the right circumstances.

However, his scientific mind reminded him that every action has consequences – something known as the butterfly effect in chaos theory which states that small changes can lead to significant effects over time or across different systems.

So what would happen if he interfered with this world? Would it cause ripples affecting their reality or even ours?

These thoughts made Sam realize the enormity of his discovery and responsibility on his shoulders. He decided then to observe without interfering until he understood more about these parallel universes - their rules, their workings.

And so began Sam's journey from being a mere observer to becoming an interdimensional explorer navigating through realities beyond human comprehension!

* * *

Now dear reader imagine yourself in Sam's shoes... What would you do if you stumbled upon such extraordinary discovery? How would you navigate your way around unfamiliar territories?

Here are some exercises for you:

1) Write down how you'd feel stepping into another universe.
    2) List three things you'd want to explore first.
    3) Think about potential challenges & ways to overcome them.
    4) Consider ethical implications of interacting with alternate realities.

Remember there are no wrong answers here! This exercise is meant to stimulate your imagination allowing us together to delve deeper into the fascinating world of multiverses!

As we continue our journey with Sam, let's keep an open mind and remember - in a universe of infinite possibilities, anything is possible!

NAVIGATING THROUGH DIFFERENT REALITIES

Chapter 11: Navigating Through Different Realities

In the previous chapters, we have explored the concept of a multiverse and our hero's first encounter with a parallel universe. Now, let us delve into how one might navigate through these different realities.

Imagine standing at the edge of an ocean, where each wave represents a separate reality within this vast multiverse. Each ripple is unique in its shape and size, just as every universe has its own set of physical laws and constants. The challenge lies not only in finding a way to cross this cosmic sea but also in understanding how to survive and thrive amidst such diverse conditions.

Our hero found himself facing this very predicament when he stumbled upon his first parallel world - an Earth where dinosaurs never went extinct. He had to quickly adapt to avoid becoming lunch for a hungry T-Rex! This experience taught him that navigating through different realities requires both mental agility and physical resilience.

Let's consider some real-world examples that can help illustrate this concept further. In business, successful entrepreneurs are often those who can navigate changing market trends effectively. They must be able to anticipate shifts in consumer behavior or technological advancements before their competitors do so they can adjust their strategies accordingly.

Similarly, survivalists train themselves to adapt quickly to various environments – from dense forests to arid deserts – by learning skills like building shelters or identifying edible plants. These individuals understand that flexibility is key when dealing with unpredictable circumstances.

Now imagine applying these principles on an interdimensional scale! Our hero would need more than just survival skills; he'd need knowledge spanning multiple scientific disciplines – physics for understanding varying laws of nature across universes; biology for recognizing alien flora & fauna; sociology for interacting with alternate societies; even philosophy for grappling with existential questions posed by encountering 'what could have been' scenarios!

So how does one prepare oneself for such mind-boggling challenges? Here are some exercises you might find helpful:

1. **Mental Flexibility Exercises**: These can range from puzzles and riddles that challenge your problem-solving skills to meditation techniques that help you stay calm under pressure.

2. **Interdisciplinary Learning**: Expand your knowledge base by exploring different fields of study. The more diverse your understanding, the better equipped you'll be to handle unexpected situations.

* * *

3. **Physical Training**: While we don't have interdimensional gyms yet, regular exercise can boost both physical stamina and mental resilience – qualities essential for any multiverse explorer!

As our hero continues his journey through parallel worlds, he learns something new from each reality he encounters - whether it's a novel way to harness energy from a binary star system or an alien philosophy that challenges his worldview. He realizes that navigating through different realities is not just about survival; it's also about growth and evolution as an individual.

In the words of physicist Niels Bohr: "How wonderful that we have met with a paradox. Now we have some hope of making progress." As daunting as this multiversal maze may seem, remember - every challenge encountered is but another opportunity for learning and growth!

So are you ready to embark on your own adventure across multiple realities? Remember, in the grand scheme of the multiverse, there are infinite paths waiting to be explored!

CAUSALITY, TIME TRAVEL, AND PARADOXES IN THE MULTIVERSE

Chapter 12: Causality, Time Travel, and Paradoxes in the Multiverse

As our hero Sam delves deeper into the labyrinth of parallel universes, he encounters a concept that challenges his understanding of reality - causality. In our everyday lives, we take for granted that cause precedes effect. You knock over a glass (cause), it falls and breaks (effect). But what if this fundamental principle doesn't hold true in all corners of the multiverse?

Let's start by defining causality. It is the relationship between an event (the cause) and a second event (the effect), where the second event is understood as a consequence of the first. However, when you venture into quantum mechanics or step through portals to other dimensions, things get tricky.

Consider time travel – one of science fiction's most beloved themes but also one fraught with paradoxical implications. Imagine traveling back in time to meet your younger self; this scenario presents us with what's known as "The Grandfather Paradox." If you were to prevent your grandparents from meeting each other before they had children, then your parents wouldn't be born...and neither would you! So how could you have existed to go back in time in the first place?

This paradox has been explored extensively throughout pop culture – think Marty McFly's precarious situation in "Back To The Future" when he accidentally interferes with his parents' meeting.

Our hero finds himself grappling with these mind-bending concepts as he navigates through different realities within the multiverse. He realizes that actions taken within these worlds can ripple across others creating unforeseen consequences - much like throwing a stone into still water creates ripples affecting even distant shores.

But let's not forget about quantum mechanics' interpretation of causality which introduces another level of complexity altogether! According to some interpretations such as Hugh Everett III's Many-Worlds Interpretation every decision made spawns new universes where each possible outcome happens. This means that causality might not be as linear as we perceive it to be.

Now, let's take a moment and think about how this concept of non-linear causality could impact our lives. Imagine if every decision you made created an alternate universe where the other choice was taken? What would your life look like in those universes?

To better understand these concepts, let's engage in a thought experiment:

Imagine you have two choices for dinner tonight - pizza or salad. In our world, you choose one and proceed with your evening. But according to the Many-Worlds Interpretation, making this choice splits reality into two parallel universes – one where you chose pizza and another where you opted for salad!

This chapter is just scratching the surface of causality, time travel, and paradoxes within the multiverse theory. As our hero continues his journey through different realities he will encounter more complex situations challenging his understanding of these principles.

In conclusion remember that while these concepts may seem far-fetched they are being seriously considered by some physicists today! So next time when faced with a difficult decision don't stress too much because somewhere out there in another universe another version of 'you' has probably chosen differently!

THE BUTTERFLY EFFECT ACROSS MULTIPLE UNIVERSES

Chapter 13: The Butterfly Effect Across Multiple Universes

In the previous chapters, we have explored the concept of parallel universes and how our hero Sam has begun to navigate through these different realities. Now, let's delve into one of the most fascinating aspects of multiverse theory - The Butterfly Effect across multiple universes.

The term "Butterfly Effect" was first coined by meteorologist Edward Lorenz in 1963 during a talk titled "Does the flap of a butterfly's wings in Brazil set off a tornado in Texas?" It refers to the idea that small changes can lead to significant effects over time. In other words, it is an embodiment of chaos theory where initial conditions play a crucial role in determining long-term outcomes.

Imagine this scenario: Sam finds himself standing at Times Square on New Year's Eve. He decides to buy a hot dog from one vendor instead of another. This seemingly insignificant choice sets off an entirely new chain reaction leading him down an alternate path he would not have taken otherwise.

Now extrapolate this concept across multiple universes – each decision creating ripples that affect not just our universe but countless others as well. Every action we take could potentially create or destroy entire worlds without us even realizing it.

Consider for instance famous historical events like World War II or man landing on moon; they might never have happened if someone somewhere had made different choices.

This raises some intriguing questions about free will and determinism - are we truly masters of our destiny? Or are we merely puppets dancing on strings pulled by cosmic forces beyond our comprehension?

Let's turn back to our hero's journey now. As he navigates through various dimensions, his actions inadvertently cause drastic changes across several parallel worlds – sometimes for better, sometimes worse.

For example, in one world where he helps someone out with their groceries, that person goes onto invent technology that revolutionizes healthcare saving millions lives; while in another world where he accidentally bumps into a stranger, it triggers a series of events leading to global catastrophe.

This is the Butterfly Effect in action across multiple universes. It's an overwhelming concept to grasp and can lead one down a rabbit hole of existential dread. But our hero learns to cope with this reality by focusing on what he can control – his actions and intentions.

He realizes that while he cannot predict or control the outcomes, he can choose to act with kindness and integrity regardless of circumstances. This realization empowers him as he continues his journey through different realities.

* * *

Now let's take this knowledge back into our own world. We may not be able to traverse parallel universes like our hero but we do make choices every day that have far-reaching consequences.

So here's an exercise for you: Think about your daily decisions - from what you eat for breakfast, how you commute work, who you interact with etc., then imagine how these could potentially ripple outwards affecting others around you and even people halfway across globe!

Remember - just like our interdimensional traveler, each one us holds power shape countless worlds through seemingly insignificant choices we make every day.

In conclusion, understanding the Butterfly Effect across multiple universes teaches us about interconnectedness all things; it reminds us that we are part larger cosmic tapestry where everything affects everything else in ways beyond human comprehension.

THE HERO'S CHALLENGE – A BATTLE ACROSS DIMENSIONS

Chapter 14: The Hero's Challenge – A Battle Across Dimensions

Our journey through the multiverse has been a thrilling ride so far, filled with mind-bending concepts and awe-inspiring revelations. But now, we've reached a pivotal point in our narrative where theory meets practice, and our hero must face his first real challenge - a battle across dimensions.

Imagine this scenario: Our protagonist, let's call him Sam Doe for simplicity's sake, is an ordinary man from Earth who stumbled upon the ability to traverse different universes. After exploring various realities and learning about the infinite possibilities that exist within them, he suddenly finds himself confronted by an interdimensional entity threatening his home universe.

* * *

The concept of battling across dimensions might seem like something straight out of science fiction or comic books. Still, it serves as an excellent metaphor for overcoming challenges that appear insurmountable at first glance. Just as Sam needs to use all his knowledge about parallel universes to protect his own world, we too often find ourselves needing to draw on all our resources and experiences when faced with significant obstacles in life.

Let's take historical examples into account; consider Winston Churchill during World War II. He was not only fighting against physical enemies but also combating despair within his nation while maintaining diplomatic relations with allies—essentially fighting battles on multiple fronts or 'dimensions.'

Similarly, Sam faces adversaries not just physically but mentally and emotionally as well. He grapples with fear over the unknown entity's power and guilt over potentially endangering other universes due to actions taken in his own reality—a classic example of cause-and-effect rippling through multiple dimensions.

To win this battle across dimensions requires more than brute force— it demands strategy, understanding one's strengths & weaknesses (both personal & dimensional), resilience under pressure & most importantly—the courage to make hard choices when necessary.

Sam begins by gathering information about this entity—its origin dimension? Its powers? Its weaknesses? He then devises a plan, using his knowledge of different dimensions to his advantage. For instance, if the entity is from a universe where time flows differently, he could use this to outmaneuver it.

This chapter serves as an exercise in problem-solving and strategic thinking for readers. It encourages you to think about how you would handle such a situation—what resources would you draw upon? How would your understanding of different 'dimensions' (whether they be physical realities or metaphorical aspects of life) aid in your strategy?

* * *

As we follow Sam's journey through this interdimensional conflict, let's also reflect on our battles across multiple 'dimensions'—be it personal growth, professional challenges or societal issues. Like Sam Doe navigating through multiverses, we too can learn to navigate our complexities by understanding them better and leveraging their unique characteristics.

In conclusion: Whether literal or metaphorical—the concept of battling across dimensions underscores the importance of adaptability & resilience when faced with daunting challenges. As our hero continues his journey through the multiverse—we hope that readers will find inspiration & courage from his story—to face their own battles head-on.

MYSTERIES UNVEILED - DARK MATTER & DARK ENERGY

Chapter 15: Mysteries Unveiled - Dark Matter & Dark Energy

As our hero Sam journeyed through the multiverse, he encountered phenomena that defied his understanding. The vastness of each universe was filled with more than just stars and galaxies; it was permeated by unseen forces that held everything together or pushed them apart. These were the enigmatic entities known as dark matter and dark energy.

Dark matter is a term used to describe an unknown type of matter hypothesized to account for a large part of the total mass in the universe. Despite its invisibility, we know it exists because of its gravitational effects on visible matter such as stars and galaxies. Imagine walking into a room where you can't see anything but can feel objects pulling you from different directions – this is what encountering dark matter would be like.

On Earth, scientists have been trying to detect these elusive particles using sophisticated detectors buried deep underground, away from cosmic rays' interference. However, despite their best efforts, direct detection remains elusive.

Sam found himself in a similar situation while navigating through one particular parallel universe dominated by dark matter's influence. He could not see it directly but felt its presence everywhere around him—like an invisible labyrinth guiding his path.

In contrast to dark matter's attractive force stands another mysterious entity: Dark Energy—a hypothetical form of energy thought to permeate all space and accelerate the expansion of the universe. It acts like an anti-gravity force pushing galaxies apart at ever-increasing speeds—an idea supported by observations showing distant galaxies moving away faster than those closer to us.

Imagine being on a ship sailing across an ocean that keeps expanding—the farther you go, the wider it becomes—that's what experiencing dark energy would feel like!

The concept might seem counterintuitive initially—after all, shouldn't gravity pull things together? But remember Einstein's famous equation $E=mc^2$? It tells us that energy has mass—and thus gravity. So, it's not entirely unreasonable to think that space itself could have energy and hence gravitational effects.

Our hero had a firsthand experience of this phenomenon in another universe where dark energy dominated. He felt an inexplicable force pushing him away from galaxies, making navigation increasingly challenging as he ventured deeper into the cosmos.

The existence of dark matter and dark energy presents profound questions about our understanding of physics. They make up approximately 95% of the total mass-energy content of the universe—yet we know so little about them! It's like being handed a book with only 5% of its pages filled—the rest are blank!

As our hero journeyed through these mysterious realms, he realized how much more there is to learn and discover. The multiverse was full of surprises waiting to be unraveled—a testament to his growing wisdom that true knowledge lies in knowing one's ignorance.

To better understand these phenomena, try this thought experiment: Imagine you're standing on a planet made entirely out of dark matter. What would you see? How would you navigate? Now imagine floating in a universe filled with just dark energy—how does it feel?

These exercises might seem abstract but remember: every great scientific discovery began as an idea—a spark ignited by curiosity and nurtured by imagination!

In your own life, consider moments when you've encountered 'dark matter' or 'dark energy' - challenges or forces unseen yet profoundly influential. How did they shape your path? Reflecting on these experiences can offer valuable insights into navigating uncertainties— in life and beyond!

As we delve deeper into the mysteries surrounding us—from quantum particles to cosmic entities—we realize that science isn't merely about finding answers; it's also about asking better questions.

THE ANTHROPIC PRINCIPLE IN COSMOLOGY

Chapter 17: The Anthropic Principle in Cosmology

The universe is a vast, complex entity that has been the subject of human curiosity for centuries. One of the most intriguing questions it poses is why it appears to be so finely tuned for life as we know it. This leads us to our discussion on the anthropic principle.

The anthropic principle, coined by physicist Brandon Carter in 1973, suggests that the universe's fundamental physical laws and constants are such because they allow observers like us to exist. In other words, if these parameters were any different, intelligent life capable of observing and studying the cosmos might not have evolved at all.

To illustrate this concept further, let's consider an analogy from pop culture - Goldilocks from "Goldilocks and The Three Bears". Just as Goldilocks preferred her porridge neither too hot nor too cold but 'just right', similarly our Universe seems to have conditions 'just right' for life as we know it.

There are two versions of this principle – weak and strong. The weak anthropic principle states that humans observe only those universes hospitable to their existence while ignoring others where conditions don't permit life. On the other hand, according to the strong version, human existence itself determines how physical reality behaves.

Our hero found himself pondering over these principles one day while navigating through a parallel universe strikingly similar yet subtly different than ours - a world where gravity was slightly stronger than what he was used to back home.

In this new realm with its altered gravitational pull, buildings were shorter and sturdier; trees grew close-knit together forming dense forests; even people appeared somewhat squatter due their bodies adapting over generations under increased pressure! It seemed like every aspect had adjusted accordingly maintaining equilibrium necessary for survival despite changes in universal constants!

This experience made him realize how delicately balanced his own universe was – just enough gravity holding everything together without crushing it, just enough light from the sun to warm the planet without burning it up.

Now, let's take a moment and think about this: What if gravity was slightly stronger in our universe? Would skyscrapers be able to stand tall or would they collapse under their own weight? How would life adapt?

The anthropic principle has its critics too. Some argue that it is more of a philosophical statement than scientific theory as it doesn't provide testable predictions. Others see it as an argument for intelligent design.

However, regardless of these debates, one cannot deny that the anthropic principle provides us with a unique perspective on our existence - we are not merely passive observers but active participants shaping reality itself!

As you journey through your day today, try observing your surroundings from this new perspective – how does everything around you contribute towards making life possible? And what role do you play in maintaining this delicate balance?

Sam continued his adventures across parallel universes with newfound respect for his home world and its perfect conditions allowing him to exist and explore infinite realities!

A GLIMPSE BEYOND OUR UNIVERSE – ENCOUNTERING ALIEN CIVILIZATIONS

Chapter 18: A Glimpse Beyond Our Universe – Encountering Alien Civilizations

As we delve deeper into the concept of multiverses, it's impossible to ignore one tantalizing possibility - the existence of alien civilizations. In this chapter, we will explore what encountering such civilizations might entail and how our understanding of life itself could be fundamentally altered.

Imagine for a moment that you are our hero, having journeyed through countless parallel universes. You've seen worlds where history took different turns, where physical laws varied subtly or dramatically from those in your home universe. Now imagine landing on an Earth-like planet orbiting around a distant star in another universe entirely. As you step out onto its surface, you see before you not just unfamiliar flora and fauna but intelligent beings who look nothing like humans.

This is no longer mere speculation; it's a reality that many scientists believe is not only possible but likely given the vastness of the multiverse. The famous Drake Equation postulates that there should be numerous technologically advanced civilizations within our own galaxy alone. Expand this idea across infinite universes with their own galaxies teeming with stars and planets, and it becomes almost inconceivable that we are alone.

The encounter with an alien civilization would undoubtedly challenge us in ways unimaginable. How do they communicate? What forms do their societies take? Do they have concepts similar to ours—of art, love, war—or something entirely different?

Consider Octavia Butler's "Xenogenesis" series as an example from pop culture exploring these themes profoundly—the Oankali aliens perceive through tentacles rather than eyes or ears; they don't have gender roles as humans understand them; their society is based on mutual symbiosis rather than competition.

In real-world science too there has been much debate about what form extraterrestrial intelligence might take—biological entities like us or perhaps artificial intelligences far beyond human comprehension? Renowned physicist Michio Kaku, in his book "The Future of Humanity," suggests that advanced civilizations might even harness the power of stars or entire galaxies.

Now let's return to Sam. As he stands on this alien world, he realizes that despite all the differences, there is a common thread—life. It may not look like life as we know it, but it thrives and evolves just as life does on Earth. This realization brings with it a profound sense of connection and humility.

For your exercise today: Imagine you are part of an inter-universal expedition encountering an alien civilization for the first time. Write down how you would communicate with them without using language? What universal symbols or concepts could be used?

This encounter reminds us that while each universe within the multiverse may be unique, they are all connected by threads of possibility and existence itself. The exploration into these parallel universes isn't just about understanding their physics—it's also about understanding ourselves better and realizing our place in this grand cosmic tapestry.

TALES FROM OTHER WORLDS – STORIES FROM ALTERNATE HISTORIES

Chapter 19: Tales from Other Worlds – Stories from Alternate Histories

In the vast expanse of the multiverse, each universe is a unique tapestry woven with threads of reality and possibility. Each thread represents an alternate history, a different outcome to events that we know too well in our own world. These are not mere hypothetical scenarios or imaginative fictions; they are realities as tangible and valid as our own.

Imagine if you will, a universe where Rome never fell. A world where Latin echoes through bustling markets, grand coliseums still host gladiatorial games, and togas remain high fashion. The Roman Empire's influence permeates every aspect of life - politics, culture, technology - shaping this parallel Earth into something both familiar yet distinctly alien.

Or consider another universe where the Industrial Revolution happened centuries earlier due to advancements in steam power during medieval times. This accelerated technological progress has led to an intriguing blend of old-world aesthetics with advanced machinery—a steampunk paradise!

Our hero Sam found himself wandering through these worlds on his journey across dimensions—each one presenting him with new challenges but also enriching him with their unique histories and cultures.

As he navigated these strange lands (or should we say 'familiar-yet-different'?), he realized how much history shapes us—our beliefs, values, aspirations—and yet how it could have taken countless other forms under slightly different circumstances.

He met figures who were historical legends in his home dimension but lived entirely different lives here—an Alexander who chose philosophy over conquests or an Einstein working as a humble patent clerk all his life without ever formulating relativity theory!

These encounters made him ponder upon the fragility and randomness of fate—the same person born into different circumstances can lead vastly divergent lives.

But what about collective destiny? He witnessed societies shaped by radically altered pasts—worlds recovering from nuclear wars that occurred during World War II or civilizations flourishing under global peace pacts established in the Middle Ages.

These tales from other worlds were not just fascinating stories for our hero; they served as mirrors reflecting his own world's triumphs and failures, its potential and follies.

As you journey with him through these alternate histories, I invite you to reflect upon your understanding of history—not as a static, immutable past but a dynamic tapestry of possibilities. How would your life change if one historical event turned out differently? What about society at large?

To help you explore this concept further, here is an exercise: Choose a significant historical event—say the discovery of fire or the signing of a major treaty—and imagine how different our world would be if that had never happened or occurred differently. Write down your thoughts and discuss them with others to gain new perspectives.

Remember, each universe in the multiverse holds countless such tales —stories waiting to be discovered by intrepid explorers like our hero...and perhaps someday by us too!

In the next chapter, we will delve deeper into navigating through different realities—a skill crucial for any aspiring interdimensional traveler!

BECOMING A MASTER NAVIGATOR THROUGH PARALLEL WORLDS

Chapter 20: Becoming a Master Navigator through Parallel Worlds

In the previous chapters, we have journeyed across dimensions and realities, exploring the vast expanse of the multiverse. We've encountered strange worlds and even stranger inhabitants. Now, it's time to take our exploration to another level - becoming a master navigator through parallel worlds.

Imagine yourself as an explorer in the Age of Discovery. You're standing on the deck of your ship, staring out at an endless sea that stretches beyond the horizon. The wind is strong; your sails are full. Your compass points towards uncharted territories filled with unknown dangers but also untold wonders.

This is how our hero felt when he first started his journey across different realities – overwhelmed yet excited by infinite possibilities.

The concept of navigating through parallel universes might seem like something straight out of science fiction or fantasy novels. However, if we consider quantum mechanics' many-worlds interpretation (as discussed in Chapter 10), every decision you make could potentially create a new universe where you made a different choice.

So how does one become adept at traversing these countless realities?

Let's delve into this fascinating topic!

Understanding Quantum Superposition

Our first step towards mastering interdimensional navigation involves understanding quantum superposition – a fundamental principle in quantum mechanics stating that any physical system (such as an electron) exists partially in all its possible states simultaneously before being observed or measured.

Consider Schrödinger's famous thought experiment involving a cat placed inside a box with radioactive material and poison gas triggered by decay particles' detection from said material. Until someone opens this box to check on poor Mr Whiskers' state (alive or dead), he exists both alive AND dead according to quantum theory! This bizarre scenario illustrates superposition perfectly.

Now imagine applying this principle not just to cats but entire universes! Each universe represents one outcome among countless others for every event occurring within it – creating a vast web of parallel realities.

Mastering Quantum Decoherence

Next, we must understand quantum decoherence – the process by which quantum systems lose their 'quantumness' and behave more like classical systems due to interactions with their environment.

In our multiverse journey, this concept is crucial as it helps us distinguish between different universes. Each universe can be considered a coherent system that has decohered from other universes due to differences in outcomes of events.

Sam learned this lesson during his first encounter with a parallel universe (Chapter 11). He realized that even though he could travel across dimensions, he couldn't change the events within them because they had already 'decohered' from his original universe.

* * *

Navigating Through Choices

The final step towards becoming an expert navigator involves understanding how choices lead to branching paths into different realities. Every decision you make creates a new branch in your life's timeline leading to another version of reality where you made an alternative choice.

This realization hit our hero hard during his battle across dimensions (Chapter 15). He understood that every action he took not only affected his own fate but also created countless alternate versions of himself living out different consequences in various worlds!

Now comes the practical part: How do we navigate through these infinite possibilities? The answer lies within us - literally! By harnessing our consciousness and intent, we can potentially influence which path we take at each crossroad, thereby choosing which reality or set of realities we experience next.

As fantastical as it sounds, some scientists propose that consciousness might play an essential role in interpreting quantum mechanics and possibly navigating multiple realities!

In conclusion, mastering navigation through parallel worlds requires understanding complex concepts such as quantum superposition and decoherence while recognizing how our choices create branching paths into different realities. It's indeed no small feat but remember - every great journey begins with one small step...or in this case - one quantum leap!

Exercises for the Reader

1. Reflect on a significant decision you made recently. Now imagine an alternate universe where you chose differently. How would that reality differ from your current one?

* * *

2. Research more about quantum superposition and decoherence to deepen your understanding of these concepts.

3. Consider how consciousness might play a role in navigating multiple realities. Do you believe our thoughts can influence which path we take at life's crossroads? Why or why not?

As Albert Einstein once said, "Reality is merely an illusion, albeit a very persistent one." So buckle up, dear reader - our journey through the multiverse has only just begun!

SOLVING CROSS-DIMENSIONAL CRISES

Chapter 21: Solving Cross-Dimensional Crises

As our hero, Sam, embarked on his journey through the multiverse, he was not just a passive observer. He found himself in situations that required quick thinking and decisive action. This chapter will delve into how Sam navigated these cross-dimensional crises and what we can learn from his experiences.

Sam's first encounter with a crisis came unexpectedly during one of his early travels to an alternate universe where technology had advanced far beyond anything seen in our own world. The society there relied heavily on artificial intelligence (AI), but their AI systems were malfunctioning, causing chaos across their world.

The concept here is 'cross-dimensional crisis management.' Just like any other crisis management situation, it requires understanding the problem at hand thoroughly before attempting to solve it. In this case, the challenge was twofold - understanding both the technological aspects of this advanced civilization and also adapting to the cultural norms of this parallel universe.

Drawing parallels from real-world examples such as global pandemics or climate change crises can help us understand better how to approach such situations. These are problems that affect all countries regardless of borders – much like a cross-dimensional crisis affects multiple universes irrespective of dimensional boundaries.

In dealing with this AI malfunctioning issue, Sam couldn't rely solely on his knowledge from Earth; he needed to adapt quickly and learn about new technologies alien to him while respecting and working within societal norms different than those he knew back home.

Let's take another example from pop culture for further clarity - remember when Tony Stark aka Iron Man faced off against Thanos in Avengers: Infinity War? Tony didn't have prior experience fighting intergalactic warlords wielding infinity stones! Yet he adapted quickly by analyzing Thanos' strengths & weaknesses while leveraging his own tech-savvy skills effectively under pressure - similar traits were required for Sam too!

Now imagine you're facing your own cross-dimensional crisis. How would you react? What skills do you think are necessary to navigate such a situation?

Sam's approach was systematic: he first sought to understand the AI systems by interacting with local engineers and scientists, then identified the root cause of the malfunction - an algorithmic error that had cascaded into larger system failures.

Once Sam understood the problem, he worked collaboratively with this universe's inhabitants to devise a solution. He drew upon his knowledge from Earth but also incorporated new information learned in this alternate reality. This highlights another key aspect of cross-dimensional crisis management – collaboration and mutual learning.

In conclusion, solving cross-dimensional crises requires adaptability, quick learning, respect for other cultures or norms (even if they're alien!), and most importantly - collaborative efforts towards finding solutions.

As we journey through our own 'multiverse' here on Earth dealing with global issues like climate change or pandemics, let's remember these lessons from Sam's adventures across parallel universes!

A KING AMONGST MANY WORLDS

Chapter 23: A King Amongst Many Worlds

In the grand tapestry of the multiverse, our hero had journeyed far and wide. He'd navigated through countless realities, each one more bewildering than the last. But now, he found himself standing at a precipice - not just in any world but as a king amongst many worlds.

The concept of kingship is deeply ingrained in human history. From ancient Egypt's pharaohs to medieval Europe's monarchs, we've always been fascinated by those who rule over us. However, being a king in this context was different; it wasn't about ruling over subjects or wielding power for personal gain.

Our protagonist had become an expert navigator of parallel universes – understanding their rules and peculiarities better than anyone else could ever hope to do so. His knowledge and experience made him akin to royalty within these realms – hence his metaphorical crown.

But what does it mean to be a "king" among many worlds? Let's delve into this intriguing notion further.

Firstly, let's consider the responsibilities that come with such status. Just like earthly kings are expected to guide their kingdoms towards prosperity while ensuring peace prevails, our hero too has obligations towards these multiple universes he traverses through regularly.

He must respect each universe's unique laws and customs while also striving for harmony between them all - no easy task considering how vastly different they can be from one another! This requires wisdom beyond measure and patience that would put even Job from biblical times to shame!

* * *

Consider King Solomon renowned for his wisdom or Emperor Ashoka known for his benevolence; they were revered because they used their authority wisely rather than oppressively. Similarly, our hero needs to use his knowledge about these various dimensions responsibly without causing harm intentionally or unintentionally due to ignorance or arrogance.

Secondly comes the challenge of maintaining balance across these numerous realities—a Herculean task indeed! Just as a king must ensure his kingdom doesn't fall into chaos, our protagonist too has to prevent any catastrophic events that could disrupt the delicate equilibrium of these parallel universes.

Imagine King Arthur and his legendary Round Table. The table symbolized equality among all knights, each having an equal say in matters of importance. This balance is what our hero strives for - ensuring no single universe dominates over others or falls into oblivion due to neglect.

But how does one maintain such balance? It's not like there's a manual lying around on "How To Be A Multiversal King"! Here's where our hero's journey so far comes handy. He draws upon his experiences from past encounters with different realities and uses them as guiding principles in this new role.

Lastly, being a king amongst many worlds also means embracing humility despite possessing immense knowledge about multiple dimensions. Remember King Midas who wished everything he touched turned to gold? His arrogance led him down a path of self-destruction when even food turned into gold at his touch!

Our hero needs to remember that while he might be akin to royalty within these realms, it doesn't make him infallible or omnipotent. He must remain humble and open-minded – always willing to learn more about the infinite mysteries the multiverse holds within its folds.

* * *

In conclusion, being a "king" among many worlds isn't just about exploring various dimensions; it's also about understanding their intricacies deeply enough so you can navigate through them responsibly without causing harm or imbalance.

It's indeed an arduous task but one that brings along unparalleled wisdom and profound insights – making every challenge faced worthwhile in the end!

So here we are readers: standing alongside our protagonist as he dons this metaphorical crown amidst countless realities. As we turn towards the next chapter of this grand adventure let us ponder: How would you rule if given such responsibility?

Remember dear reader: With great power comes great responsibility. And in the multiverse, that power is not just to rule but to understand, respect and maintain harmony among countless realities.

ETERNAL RECURRENCE - NIETZSCHE'S THOUGHT EXPERIMENT

Chapter 24: Eternal Recurrence - Nietzsche's Thought Experiment

In the grand scheme of our multiverse exploration, we've encountered countless realities and infinite possibilities. But now, let us pause for a moment to ponder upon an intriguing philosophical concept that has been proposed by one of the greatest thinkers in human history – Friedrich Nietzsche. This chapter will delve into his thought experiment known as "Eternal Recurrence."

Nietzsche was not a physicist or cosmologist but rather a philosopher who dared to challenge conventional thinking and societal norms. His idea of eternal recurrence is more metaphysical than physical, yet it resonates with our journey through multiple universes.

The concept is simple yet profound: everything recurs eternally, without any variation. Every joy and sorrow, every triumph and failure you have experienced will repeat itself infinitely across time. Now imagine if this were true in the context of parallel universes; wouldn't that be mind-boggling?

Let's take our hero Sam as an example here. He started off as an ordinary individual unaware of the existence of other dimensions until he stumbled upon them accidentally (or perhaps fatefully). Since then, he has traversed numerous worlds, faced unimaginable challenges, met diverse beings from different realities – all these experiences shaping him into a seasoned navigator through parallel worlds.

Now consider Nietzsche's proposition: what if all these events are destined to recur over and over again? What if our hero is doomed (or blessed) to relive his adventures endlessly? How would this affect his perception towards life? Would it lead him towards despair knowing that nothing truly changes or would it inspire him to live each moment fully aware that it will return eternally?

This brings us back to Nietzsche's original intention behind this thought experiment which serves as both existential dread and liberating affirmation. He asked readers whether they would embrace such reality if some day or night a demon were to steal after you into your loneliest loneliness and say to you: "This life as you now live it and have lived it, you will have to live once more and innumerable times more; would you not throw yourself down and gnash your teeth and curse the demon who spoke thus?"

Nietzsche believed that those who could affirm such existence are the ones who truly love life. They are capable of finding joy in every moment, no matter how mundane or painful because they understand that these moments constitute their unique journey through existence.

In our hero's context, this might mean embracing each challenge he encounters across different dimensions as an integral part of his cosmic voyage. It means cherishing every interaction with alien civilizations knowing that these experiences contribute to his growth as a multiverse traveler.

* * *

But what about us? How do we apply Nietzsche's thought experiment in our lives? Perhaps by acknowledging that whether or not eternal recurrence is real, the idea itself can serve as a powerful tool for self-reflection. It encourages us to evaluate our actions, decisions, relationships – everything that defines our human experience.

As an exercise for readers: imagine if your life were to recur eternally without any changes - would you be contented with it? If not, what aspects would you want to change?

The concept of eternal recurrence may seem daunting at first glance but when viewed from another perspective (or perhaps from another universe), it can provide profound insights into how we perceive ourselves within this vast cosmos. After all, isn't exploring parallel worlds also about understanding ourselves better?

So let's continue on this fascinating journey through multiple universes while keeping Nietzsche's thought experiment in mind – reminding us always of the significance of each moment we encounter along the way.

HAWKING'S NO BOUNDARY PROPOSAL

Chapter 25: Hawking's No Boundary Proposal

In our journey through the multiverse, we've encountered countless wonders and mysteries. We've seen worlds that mirror our own with uncanny precision and others so alien they defy comprehension. But now, we come to a concept that is perhaps one of the most mind-bending yet - Stephen Hawking's No Boundary Proposal.

The late Professor Stephen Hawking was not just an iconic figure in popular culture; he was also one of the greatest scientific minds of our time. His work on black holes and quantum mechanics has shaped much of modern cosmology. One such contribution is his "No

Boundary Proposal," or Hartle-Hawking state, which he developed alongside American physicist James Hartle.

Imagine you're standing at Earth's North Pole. If someone asks you to move north, you'd be puzzled because there isn't any 'north' from this point – every direction would lead south! This analogy helps us understand what Hawking proposed about the universe: it might have no boundary – no beginning nor end.

Hawking suggested that if we were to rewind time back towards the Big Bang, instead of reaching a singularity—a point where physical laws break down—we would find that space-time becomes increasingly curved until it smoothly closes off into nothingness— much like how Earth closes off when moving northward till North Pole.

Sam found himself pondering over this idea as he navigated through another parallel world—one where humanity had mastered interstellar travel but still grappled with these fundamental questions about their universe's origin. He wondered whether understanding concepts like these could help him navigate better across dimensions or even manipulate them?

To truly grasp this concept, let's delve deeper into its implications:

1) **Timelessness:** In a closed-off model without boundaries (like surface of Earth), there wouldn't be any singularities for time to begin from or end at—it would become another direction in space. This implies a universe that doesn't have a distinct beginning or end, challenging our conventional understanding of time.

2) **Quantum Mechanics and General Relativity:** The No Boundary Proposal is an attempt to reconcile these two fundamental theories of physics. At the singularity point, both theories give conflicting predictions; however, if there's no singularity (as suggested by this proposal), they could potentially be unified.

* * *

3) **Predictability:** If the universe indeed has no boundaries, it would mean that its state at any given moment can be calculated from its state at any other moment—providing determinism across all of space-time!

Now let's try an exercise: Imagine you're standing on Earth's North Pole again. Visualize moving southward – every step taking you through different points in time instead of space. How does your perception change when time becomes just another direction?

Our hero found himself applying this new perspective as he navigated through his next adventure—a world where time flowed differently based on one's location! He realized how such abstract concepts weren't merely academic but had profound implications even for his journey across multiverses.

As we continue exploring parallel worlds and their myriad possibilities, Hawking's No Boundary Proposal serves as a reminder that our understanding of the cosmos is still evolving. It challenges us to rethink our notions about time and existence itself—an essential lesson for anyone venturing into unknown dimensions!

In conclusion, remember what Stephen Hawking once said: "The greatest enemy of knowledge is not ignorance; it is the illusion of knowledge." As we delve deeper into the mysteries of multiverse theory and beyond, let us keep questioning and learning with open minds.

TEGMARK'S FOUR LEVELS

Chapter 26: Tegmark's Four Levels

As Sam continues his journey through the multiverse, he stumbles upon a concept that seems to encapsulate all of his experiences and observations so far. This is the theory proposed by Max Tegmark, a renowned physicist known for his work on cosmology. The theory suggests that there are four distinct levels of parallel universes or multiverses.

The first level is what we've been exploring throughout this book - the idea of infinite space-time regions in our own universe. These are areas beyond what we can observe from Earth due to limitations in light speed and cosmic inflation. In these regions, it's possible that everything repeats eventually because there are only so many ways particles can be arranged within a finite volume.

Our hero reflects on how this concept resonates with some of his own encounters across different dimensions. He recalls stepping into worlds eerily similar to ours but with slight variations - like an Earth where dinosaurs never went extinct or another where humanity had already colonized Mars.

Next comes Level II Multiverse which introduces other universes with potentially different physical laws born out of their unique Big Bangs during cosmic inflation periods. Our protagonist remembers visiting realms where gravity was weaker or time flowed differently – manifestations of varying physical constants.

Level III Multiverse delves deeper into quantum mechanics' Many-Worlds Interpretation (MWI). It proposes every quantum event spawns new universes where each outcome happens separately – essentially creating copies of ourselves living out all possible decisions simultaneously! Our hero shudders at the thought; he'd seen doppelgängers leading lives divergent from his own based on choices they made differently!

Finally, Level IV Multiverse encompasses mathematical structures as separate realities – an abstract yet profound notion suggesting any mathematically describable universe exists physically somewhere! This level challenges conventional understanding about reality itself and leaves even our seasoned explorer awestruck!

Tegmark's theory, while controversial, provides a comprehensive framework for understanding the multiverse. It also raises profound philosophical questions about existence and reality that our hero grapples with.

Exercise: Imagine you're in a Level III Multiverse where every decision creates an alternate universe. Write down three significant decisions you've made today and imagine how different your life could be if you had chosen differently.

As our protagonist ponders over these concepts, he feels both humbled by the vastness of existence and empowered by his ability to navigate it. He realizes that even as a king among many worlds, there is still so much more to learn and explore - an endless journey through Tegmark's four levels of parallel universes.

DREAMING UP NEW REALITIES

Chapter 27: Dreaming Up New Realities

In the vast expanse of the multiverse, Sam has journeyed through countless dimensions and parallel universes. He's encountered worlds that mirrored his own with slight variations, and others so drastically different they defied comprehension. As he navigates these diverse realities, a question begins to form in his mind - can we dream up new realities? Can our thoughts and dreams shape or even create alternate universes?

The concept might seem far-fetched at first glance. After all, how could mere human thought influence something as grandiose as the fabric of reality itself? Yet when we delve into quantum mechanics' peculiar world – where particles exist in multiple states simultaneously until observed – this idea doesn't seem quite so outlandish.

Consider Schrödinger's cat experiment, where a cat inside a box is both alive and dead until someone opens it to observe its state. This paradoxical situation illustrates superposition - an essential principle in quantum physics suggesting that physical systems can exist in multiple states simultaneously until measured or observed.

Now imagine applying this concept on a universal scale. Could each decision we make cause reality to split into numerous branches representing every possible outcome? If so, then perhaps our thoughts do indeed have some power over shaping reality.

Let's take an example from pop culture for illustration purposes; think about Christopher Nolan's movie "Inception". In this film, characters are able to enter people's dreams and plant ideas within their subconscious minds which later manifest themselves into their waking lives' realities.

While "Inception" remains firmly rooted in science fiction territory (for now), it does raise intriguing questions about consciousness' role within the multiverse theory framework. Is there more interplay between consciousness and reality than what meets the eye?

To explore this further let's consider another scientific theory known as Orchestrated Objective Reduction (Orch-OR). Proposed by physicist Roger Penrose and anesthesiologist Stuart Hameroff, this theory suggests that consciousness originates at the quantum level within brain cells' microtubules. If true, it could imply a direct connection between human consciousness and the quantum world.

Our hero ponders these concepts as he traverses through different realities. He wonders if his thoughts are merely passive observers or active participants in shaping reality. Could dreaming up new realities be more than just metaphorical?

As you read this chapter, I invite you to ponder these questions too. Reflect on your own dreams and aspirations – do they hold the power to shape your reality? Or even create new ones?

To delve deeper into this concept, try out this exercise: Spend some time each day visualizing a specific goal or dream of yours in vivid detail. Imagine every aspect of it - how it looks, feels, sounds... everything! Do this consistently for several weeks while also taking concrete steps towards achieving that goal in your current reality.

Then observe any changes or synchronicities occurring around you related to your visualization practice. Are there signs indicating that your desired reality is manifesting itself into existence? Remember - science may not fully understand yet how our thoughts interact with reality but don't let that stop you from exploring its potential!

In conclusion, whether we can truly dream up new realities remains uncertain within our current scientific understanding scope. However, what's clear is that our journey across the multiverse continues to challenge us by presenting mind-bending possibilities about our universe's nature and perhaps even ourselves.

PONDERING UPON CONSCIOUSNESS & REALITY

Chapter 28: Pondering Upon Consciousness & Reality

In our journey through the multiverse, we have encountered many strange and wondrous things. We've seen worlds where time runs backward, universes composed entirely of antimatter, and realities where the laws of physics as we know them are turned on their heads. But perhaps one of the most profound questions that arise from these explorations is not about what's out there in those other universes— it's about us.

What does it mean to be conscious? What is reality?

Our hero Sam has been grappling with this question ever since his first encounter with a parallel universe. He had always assumed that consciousness was something unique to him—that he was an individual observer experiencing a singular reality. But his travels across dimensions have challenged this assumption.

* * *

Consider for a moment how you perceive your surroundings. You see objects around you because light bounces off them and enters your eyes; you hear sounds because vibrations travel through air or water into your ears; you feel textures because nerve endings in your skin respond to pressure.

But all these sensory inputs are just information—data being fed into your brain which then processes it and creates an experience—a representation of reality inside your head. This realization can be both liberating and disconcerting at once.

Now imagine if there were multiple versions of 'you' spread across different universes, each receiving slightly different sensory data based on their respective environments—would they all construct the same version of reality? Or would each 'you' live within its own unique interpretation?

This thought experiment brings us face-to-face with some deep philosophical questions: Is there such thing as objective truth? Or is everything subjective—dependent upon who or what is perceiving it?

The famous philosopher Immanuel Kant argued that while we may never truly access things as they are in themselves (what he called "noumena"), we can only interact with how they appear to us (the "phenomena"). In other words, our consciousness shapes our reality.

But what if we could share experiences across different versions of ourselves in various universes? Could we then create a more 'complete' picture of reality?

Our hero had an opportunity to explore this concept when he encountered a civilization that had developed technology allowing them to link minds not just within their own universe but across multiple realities. This society lived in a state of shared consciousness, experiencing life through countless perspectives simultaneously.

The experience was overwhelming at first—like trying to listen to every single instrument in an orchestra playing all at once—but over time, he learned how to tune into individual streams of consciousness and even contribute his own experiences into the collective pool.

This encounter forced him to reevaluate his understanding of selfhood. If he could access memories and thoughts from countless other selves, where did 'he' end and others begin? Was there even such thing as a singular 'self'? Or were they all just facets of some greater cosmic consciousness spread out across the multiverse?

As you ponder these questions yourself, consider this exercise: Try spending few minutes each day imagining what it would be like if there were another version of you living in a parallel universe. What kind of life might they lead? How would their world differ from yours? And most importantly—how would their experiences shape their perception of reality?

In doing so, perhaps you'll gain new insights about your own existence—and who knows—you may even start seeing your everyday world in a whole new light!

Remember—the journey through the multiverse is not just about exploring outer space—it's also about delving into inner space—the vast landscape within our minds. So buckle up for the ride because things are about to get even more interesting!

LAYING DOWN THE CROWN – RETURNING HOME

Chapter 29: Laying Down the Crown – Returning Home

In every hero's journey, there comes a time when he must lay down his crown and return home. It is not a moment of defeat but rather one of triumph, for it signifies that our protagonist has completed his mission and learned valuable lessons from his adventures in the multiverse.

Our hero had been to countless worlds, each with its unique laws of physics, strange creatures, and diverse cultures. He had seen civilizations rise and fall; witnessed stars being born and dying out; battled interdimensional beings; navigated through paradoxes that twisted logic into pretzels. But now it was time to return home.

The concept of 'home' can be quite abstract when you've traversed multiple realities. Is it merely your point of origin? Or does it hold deeper meaning - perhaps an emotional connection or sense of belonging? For our hero, home was where he started this incredible journey - Earth.

As he prepared for his departure from the last parallel universe on his itinerary - a world where gravity worked sideways (quite literally) - he felt an odd mixture of relief and melancholy. The thrill of exploration was addictive indeed! Yet there was also comfort in familiarity – the taste of mom's apple pie waiting back at home seemed more appealing than ever!

Before leaving though, our hero took one last look around him. This world had taught him about resilience as he watched its inhabitants adapt to their unusual environment with grace and ingenuity. They built structures horizontally instead vertically due to their peculiar gravitational pull—something architects back on Earth would find mind-boggling!

This experience reminded him once again how diverse yet interconnected all universes were within this grand multiverse tapestry—a lesson worth sharing with fellow earthlings upon returning.

With these thoughts swirling in his head like galaxies spinning in space-time continuum fabric itself—he stepped onto the portal leading back home.

Upon arrival on Earth, he felt an overwhelming sense of nostalgia. The familiar sights and sounds filled him with warmth as he walked down the streets of his hometown. He was a king who had laid down his crown, but in doing so, he had gained something far more valuable - wisdom.

He realized that every universe he visited was like a mirror reflecting different aspects of himself. Each adventure challenged him to grow and adapt; each encounter with alien species broadened his perspective on life; each paradox unraveled revealed new layers of understanding about reality itself.

As our hero settled back into his earthly existence, he found himself changed profoundly. His experiences across multiple dimensions made him see Earth through fresh eyes—appreciating its beauty and fragility even more.

But how could one integrate such extraordinary experiences into ordinary life? How would you share stories about parallel universes at dinner parties without sounding insane?

Our hero decided to use metaphors and analogies from pop culture to explain complex multiverse concepts subtly—he became an author! His books were filled with vivid descriptions of otherworldly adventures while subtly educating readers about quantum mechanics, string theory, cosmic inflation—all disguised as science fiction!

In this way, our protagonist continued sharing the knowledge acquired during his interdimensional travels—inspiring others to look beyond their immediate realities towards infinite possibilities lying within the multiverse's expanse.

So here's your exercise for today: Imagine yourself returning home after a long journey across multiple universes. What lessons have you learned? How have these experiences changed your perception about 'home'? And most importantly – how would you share these insights with those around you?

Remember: Every end is just another beginning in disguise—and so it was for our hero too! As we close this chapter on 'Laying Down the Crown', remember that there are still countless worlds waiting out there... beckoning us towards new adventures!

INTEGRATING EXPERIENCES FROM COUNTLESS WORLDS

Chapter 30: Integrating Experiences from Countless Worlds

Our hero, after having journeyed through a myriad of parallel universes and experiencing realities beyond human comprehension, now faces the daunting task of integrating these experiences into his own reality. This chapter will explore how one can assimilate such profound knowledge and wisdom gained from countless worlds.

Imagine for a moment that you've just returned home after an extensive trip around the world. You've seen cultures vastly different from your own, tasted exotic foods, heard languages you didn't understand, and witnessed breathtaking landscapes that exist outside your everyday experience. Now imagine magnifying those feelings exponentially - this is what our hero must grapple with.

The first step in integration is understanding. Sam has to make sense of all he's seen and experienced across multiple dimensions. He needs to process each memory individually while also considering them collectively as part of his multiverse journey.

Consider Albert Einstein's theory of relativity; it was not merely an intellectual exercise but a radical shift in perspective on space-time continuum itself. Similarly, our hero must reconcile his new understanding with previously held beliefs about reality.

Next comes acceptance – accepting that although some experiences were terrifyingly alien or overwhelmingly beautiful, they are all part of the vast tapestry of existence within the multiverse. Just like Neil Armstrong had to accept walking on moon as reality despite its surreal nature.

Then there's adaptation – applying lessons learned from other worlds into daily life here on Earth. For instance, if our protagonist observed a parallel universe where renewable energy sources have completely replaced fossil fuels leading to cleaner environment and healthier lives; could he champion for more sustainable practices in his own world?

This brings us to sharing – recounting tales from distant realms not only enlightens others but also helps solidify one's own understanding by putting abstract concepts into words.

Finally comes growth - using these integrated experiences for personal development and evolution.

Let's take Thomas Edison for example. His relentless pursuit of knowledge and countless experiments led to inventions that revolutionized our world. Similarly, our hero's experiences across the multiverse can lead to groundbreaking insights about existence itself.

Now, let's consider a practical exercise: Imagine you've just returned from an alternate reality where peace reigns supreme due to absence of any form of conflict. How would this experience change your perspective on conflicts in your own life? Would it inspire you to seek peaceful resolutions more actively?

Our hero is no longer the same person who embarked on this journey; he has grown into a king amongst many worlds, carrying within him wisdom from countless realities. He understands now that every universe is but a reflection of infinite possibilities and each one holds lessons for those daring enough to venture beyond their known realms.

In conclusion, integrating experiences from parallel universes isn't merely about understanding or accepting new concepts; it's about evolving as an individual and potentially as a species by learning from these boundless cosmic libraries of knowledge.

As Carl Sagan once said, "Imagination will often carry us to worlds that never were. But without it we go nowhere." So too must our hero - and indeed all explorers venturing into the multiverse - use their imagination not only as a vehicle for exploration but also as tool for integration upon return.

IMPLICATIONS FOR HUMANITY – LESSONS FROM THE MULTIVERSE

Chapter 31: Implications for Humanity – Lessons from the Multiverse

In our journey through the multiverse, we have encountered numerous realities and dimensions. We've seen worlds where history took a different turn, where physics operates under different laws, and even places where life as we know it doesn't exist at all. These experiences can be overwhelming, but they also offer us invaluable lessons about ourselves and our place in the cosmos.

The first lesson is humility. The vastness of the multiverse reminds us that we are just one small part of an infinitely complex tapestry of existence. Our planet, our solar system, even our entire universe might be nothing more than a tiny bubble in an endless cosmic ocean.

This realization may seem daunting or even frightening at first glance. But upon deeper reflection, it's actually quite liberating. It means that every action we take has infinite potential to create ripples across countless realities.

Consider this hypothetical scenario: In another universe very similar to ours, you decided to become a doctor instead of pursuing your current career path. You saved lives and made significant contributions to medical science there - achievements that you could never dream of in this reality.

* * *

Does this make your accomplishments here any less valuable? Absolutely not! Because each decision you make creates a new branch on your personal timeline tree spreading across multiple universes.

Our hero learned this lesson during his adventures when he met versions of himself who had taken different paths in life - some were kings while others were slaves; some lived fulfilling lives while others struggled with regret and despair.

But no matter how their circumstances varied across these parallel universes, they all shared one thing in common: They were all him - shaped by their choices and experiences into unique individuals yet fundamentally connected by their shared origin.

This leads us to another important implication for humanity: The power of choice.

Every decision matters because each choice leads down a different path creating alternate realities branching out endlessly. This is a powerful reminder of our agency and the impact we can have on our own lives and those around us.

But what about fate? If there are infinite versions of ourselves living out every possible outcome, does that mean our destinies are pre-determined?

Not necessarily. The multiverse theory doesn't negate free will; it merely expands its scope across multiple dimensions. We still make choices in this universe - choices that shape not only our reality but countless others as well.

This brings us to the final lesson: Responsibility.

With great power comes great responsibility, as the saying goes. And if we indeed possess such immense power over the fabric of existence itself, then we must wield it wisely.

Our hero learned this when he had to resolve a cross-dimensional crisis threatening his home universe. He realized that his actions could potentially affect an infinite number of realities, and so he acted with caution and consideration for all possible outcomes.

In conclusion, exploring the multiverse offers profound insights into human nature and our place in existence. It teaches us humility by reminding us how small we are compared to the vastness of infinity; it empowers us by showing how each choice creates new realities; and finally, it instills in us a sense of responsibility towards ourselves and other beings spread across countless universes.

As Carl Sagan once said: "We're made of star stuff." But perhaps more accurately - we're made up not just from matter forged in stars within one universe but also from possibilities scattered across endless parallel universes waiting for exploration.

THE HERO'S FINAL REFLECTIONS

Chapter 32: The Hero's Final Reflections

As Sam sat on the edge of a cliff overlooking an alien sunset, he couldn't help but marvel at the journey he had undertaken. He was no longer just a man from Earth; he was now a seasoned traveler of parallel universes, having witnessed realities beyond his wildest dreams and darkest nightmares.

He remembered how it all started with an unexpected discovery that led him to explore infinite universes. Each universe presented its own set of challenges and wonders, pushing him to adapt and grow in ways he never thought possible. He had seen worlds where time flowed backward, where gravity pulled sideways, where life evolved into forms unimaginable by earthly standards.

The hero reflected on his encounters with different versions of himself - some leading lives remarkably similar to his own while others were starkly different. These interactions forced him to confront uncomfortable truths about himself and question what truly defined him as an individual.

He pondered over the concept of free will versus determinism across multiple realities. Did every decision lead to another universe being created? Or were there predetermined paths laid out for each version of himself?

His adventures also brought forth ethical dilemmas that challenged his moral compass. How does one decide what's right or wrong when dealing with alternate histories or potential futures?

In one particular universe, he fell in love – a deep connection transcending dimensions which made him realize that emotions like love could exist even amidst chaos and uncertainty.

However, not all experiences were pleasant ones. There were times when fear gripped him as unknown entities lurked around corners or existential crises loomed large due to mind-bending paradoxes associated with multiverse theory.

Yet through it all, hope remained constant – hope for understanding the mysteries surrounding dark matter & dark energy; hope for encountering advanced civilizations who might have answers; hope for humanity's future in this vast multiverse.

Sam realized how these experiences shaped not just his understanding of the multiverse, but also his perspective on life. He learned to appreciate the beauty in diversity and chaos, understood the importance of resilience and adaptability, recognized the power of choice even in a deterministic universe.

He thought about how these lessons could be applied back home - to foster inclusivity by embracing differences; to encourage curiosity and exploration beyond our comfort zones; to promote sustainability by learning from civilizations that had successfully harnessed their resources without self-destruction.

* * *

As he watched alien stars twinkle against an indigo sky, Sam felt a sense of peace wash over him. His journey through parallel worlds had been more than just physical travel – it was a voyage into the depths of his soul, challenging preconceived notions and expanding horizons.

His final reflections were interrupted as he noticed something unusual in the night sky - a new constellation forming or perhaps another universe beckoning? With renewed excitement, he prepared for yet another adventure because for explorers like him, every end is simply a new beginning...

In this vast multiverse filled with infinite possibilities and endless adventures waiting around every corner... who knows what tomorrow might bring?

EPILOGUE - THE ENDLESS JOURNEY CONTINUES

Chapter 33: Epilogue - The Endless Journey Continues

As we stand on the precipice of understanding, gazing out into the infinite possibilities that stretch before us in this vast multiverse, it's important to remember one thing. Our journey is far from over.

The hero of our story has traveled through countless parallel universes, faced unimaginable challenges and returned home a king amongst many worlds. He has seen realities where he was a pauper, a prince, an artist, a scientist; realities where he never existed at all. Each universe presented its own unique set of rules and circumstances but also held up a mirror to his own existence.

In each world he visited, he found pieces of himself scattered across dimensions like cosmic breadcrumbs leading him back home. And yet with every return trip to his original universe – our universe – there was always something new waiting for him on the other side.

* * *

This endless cycle of discovery and self-reflection is not just limited to Sam's journey but extends to us as well - readers who have accompanied him throughout this grand adventure across multiple realities.

We've learned about quantum mechanics and string theory; explored concepts such as causality and determinism; pondered upon consciousness & reality; grappled with ethical dilemmas posed by life in infinite realities. We've been introduced to theories that challenge our understanding of time and space itself while simultaneously expanding our horizons beyond what we thought possible.

But perhaps most importantly, we've learned how even amidst chaos or uncertainty—whether it be navigating through different dimensions or facing unknown entities—we can still nurture hope and face fear head-on.

Our exploration doesn't end here though—it merely changes direction. As physicist John Wheeler once said: "We live on an island surrounded by a sea of ignorance." But every question answered leads us further inland towards knowledge while simultaneously revealing more questions hidden beneath the waves around us—a testament to both human curiosity's insatiable nature and the universe's infinite complexity.

So, what does this mean for us? It means that our journey through the multiverse is far from over. Each day brings new discoveries, each moment presents a chance to learn something new about ourselves and the world around us. We are not just observers in this grand cosmic play but active participants shaping our own narratives.

As we close this chapter of our exploration into parallel universes, let's remember that every end is merely a new beginning in disguise. The hero's journey continues as does ours—into uncharted territories, towards inclusive cosmology; evolving as an intergalactic species while integrating knowledge from multiple disciplines.

* * *

In conclusion, whether you're a scientist seeking answers to profound questions or simply someone fascinated by the concept of parallel worlds—the endless journey through the multiverse awaits you with open arms. So here's to curiosity—to daring dreams and audacious adventures! Here's to exploring infinity!

Remember: In a multiverse where anything can happen—you are limited only by your imagination.

THE PHYSICS OF PARALLEL UNIVERSES

Chapter 34: The Physics of Parallel Universes

In the grand tapestry of existence, our universe might be just one patch in a quilt that stretches out infinitely. This is not science fiction; this is the realm of theoretical physics and cosmology. Welcome to Chapter 34, where we delve into the fascinating world of parallel universes.

The concept of parallel universes has been around for decades, but it's only recently that physicists have started taking it seriously as a viable scientific theory. It all starts with quantum mechanics - the branch of physics dealing with particles at their most fundamental level.

Quantum mechanics introduces us to an array of strange phenomena like superposition (where particles can exist in multiple states simultaneously) and entanglement (where two particles become linked so that changes to one instantly affect the other). These concepts are challenging enough within our own universe, but they also open up tantalizing possibilities when applied on a cosmic scale.

Consider Schrödinger's famous thought experiment featuring his unfortunate cat which exists in a state of being both alive and dead until observed. Now imagine applying this idea to entire universes – each existing in different states until observed or interacted with. This leads us towards Hugh Everett's Many-Worlds Interpretation (MWI).

Everett proposed that every time there's more than one possible outcome, all outcomes occur – each within its own separate universe. So if you flip a coin, there's another version of you who experiences heads while you experience tails!

But how does this relate to actual physical reality? Enter string theory - another cornerstone in understanding parallel universes.

String theory suggests that everything we perceive as matter is made up from tiny strings vibrating at different frequencies. But here's where things get interesting: String Theory requires extra dimensions beyond our familiar three spatial ones plus time!

These additional dimensions could provide room for other 'brane-worlds' or parallel universes existing alongside our own. Each universe could be a separate 3D brane floating in higher-dimensional space.

Sam, having journeyed through multiple universes and witnessed their wonders, now faces the daunting task of understanding the physics that governs these worlds. He is like an explorer trying to map out uncharted territories using only his intuition and rudimentary tools.

As he grapples with quantum mechanics and string theory, he realizes that these theories are not just abstract concepts but have real-world implications. They challenge our fundamental understanding of reality and force us to reconsider what we think we know about time, space, matter - even existence itself!

The idea of parallel universes also raises profound philosophical questions: If there are infinite versions of ourselves living out different lives in parallel universes, what does this mean for free will? Are we simply following one path among many predestined routes?

* * *

To help you explore these ideas further here's a thought experiment: Imagine waking up tomorrow in a world where everything is identical except for one small detail – perhaps your favorite book never existed or your best friend has no memory of you. How would this change your perception of reality? Would it make your experiences any less valid?

In conclusion, while the physics behind parallel universes may seem complex (and indeed it is), they offer us new ways to understand our place within the cosmos. As our hero continues his journey across dimensions, let's join him by opening our minds to the possibilities that lie beyond our known universe.

PHILOSOPHICAL IMPLICATIONS OF THE MULTIVERSE

Chapter 35: Philosophical Implications of the Multiverse

In our journey through parallel universes and multiverses, we have encountered a myriad of realities that challenge our understanding of existence. The exploration has not only been scientific but also deeply philosophical. As we delve into this chapter, let's ponder upon some profound questions that arise from the concept of the multiverse.

The first question is about reality itself - what does it mean to be real? If there are infinite versions of ourselves living out different lives in parallel universes, which one is the 'real' us? Is reality merely a construct based on our perception?

Consider our hero who started as an ordinary individual in his universe but became a king amongst many worlds. Each version was as real to him as any other when he experienced them. This brings us to Heraclitus's famous quote "No man ever steps in the same river twice," suggesting that change is the only constant and everything else is transient.

Next comes determinism versus free will – if every possible outcome exists somewhere in the multiverse, do we truly have free will or are all choices predetermined? Our hero made decisions throughout his journey; however, each choice led him down a path already existing within another universe.

This leads us to contemplate morality within such an expansive framework. Are actions morally right or wrong across all universes or does morality shift with changing realities? When faced with ethical dilemmas during cross-dimensional crises, our hero had to adapt his moral compass according to each world's unique circumstances.

Another intriguing aspect involves personal identity and selfhood. How do we define 'self' amidst countless duplicates spread across multiple dimensions? Despite encountering numerous versions of himself throughout various worlds, each iteration maintained its distinct sense of selfhood due to differing experiences and memories formed along their respective paths.

Moreover, how should one cope with existential dread knowing they're just one among infinite selves scattered across endless realities? Our hero, initially overwhelmed by this realization, eventually found solace in understanding that each version of him was unique and had its own purpose.

The concept of the multiverse also challenges our perception of time. If time flows differently across various universes, is it a linear construct or merely another dimension we traverse? This echoes Einstein's words: "Time is an illusion."

Finally, what does death mean within the context of infinite realities? Is there truly an end or just a transition into another universe?

As we ponder these philosophical implications, let's remember that they are not meant to unsettle us but rather expand our horizons. Just like our hero who started as a mere individual but grew into a king amongst many worlds through his journey across dimensions.

* * *

Exercise for readers: Reflect on how your perspective about selfhood and reality would change if you discovered parallel versions of yourself existing simultaneously in different universes. Write down your thoughts and discuss them with others to gain diverse insights.

In conclusion, while the multiverse theory opens up new frontiers for scientific exploration, it equally enriches philosophical discourse by challenging traditional notions about reality, identity, morality and existence itself.

ETHICAL DILEMMAS IN A MULTIVERSE

Chapter 36: Ethical Dilemmas in a Multiverse

As our hero, now a seasoned traveler of the multiverse, continues his journey across dimensions, he encounters an aspect of this infinite reality that is as complex and challenging as the scientific theories themselves - ethics. The concept of right and wrong, moral and immoral takes on new meanings when applied to an infinite number of realities.

Let's start with a simple question: If there are infinite versions of you living out every possible life scenario in parallel universes, what does it mean for your sense of self? Does it dilute personal responsibility or amplify it?

Consider this hypothetical situation. In one universe, you might be a philanthropist contributing significantly to society's betterment. In another universe, however, you could be leading a criminal organization causing harm to many people. Are these actions mutually exclusive or do they somehow balance each other out across multiple realities?

This brings us back to our hero who has seen himself in various roles during his travels through different dimensions – from being a savior in one world to becoming the cause for destruction in another. How should he reconcile with these diverse identities? Is there any universal ethical code applicable throughout the multiverse?

Now let's delve into another conundrum - if we have access to alternate realities where all possibilities exist simultaneously; can we justify manipulating outcomes by interfering with events happening elsewhere? For instance, if our hero learns about an impending disaster threatening Earth-2 (a parallel version of our planet), would intervening be considered heroic or intrusive meddling?

The famous "Trolley Problem" thought experiment illustrates this dilemma well. Imagine standing at switch points controlling which track an approaching trolley will take. On its current course (track A), five people tied up will get hit; but if switched onto track B instead only one person gets hit who otherwise wouldn't have been harmed had no action been taken.

In traditional ethics, the utilitarian approach would advocate for switching to track B, thereby minimizing overall harm. However, in a multiverse context where every possible outcome exists simultaneously (one universe where you switch tracks and another where you don't), does this decision still hold moral weight?

Furthermore, if we accept that our actions can have ripple effects across multiple universes (the butterfly effect), then even seemingly insignificant choices could lead to catastrophic consequences elsewhere. This raises questions about free will and determinism - are we morally responsible for all potential outcomes of our decisions or only those within our immediate perception?

Our hero grapples with these ethical dilemmas as he navigates through the multiverse. He realizes that his journey is not just about exploring different realities but also understanding the profound implications it has on morality.

In conclusion, while science may provide us with tools to explore parallel universes theoretically or perhaps practically someday; it's philosophy and ethics that guide us on how to navigate these infinite possibilities responsibly.

As an exercise for readers: Reflect upon your own ethical beliefs. How might they change if confronted with the reality of a multiverse? Would certain actions become more justified or less so? Consider writing down your thoughts and discussing them with others – after all, pondering such complex issues is part of what makes us human.

Remember: In this vast cosmos filled with endless possibilities, each choice matters - not just here and now but potentially across countless other realities too.

LIFE AND DEATH IN INFINITE REALITIES

Chapter 37: Life and Death in Infinite Realities

In the vast expanse of the multiverse, where countless realities coexist simultaneously, life and death take on a whole new meaning. The concept of mortality becomes as fluid as the boundaries between dimensions.

Imagine our protagonist, Sam Doe - an ordinary man turned interdimensional explorer. He has journeyed through numerous parallel universes, each with its unique laws of physics and biology. In some worlds, he encountered versions of himself that had lived vastly different lives; in others, he found societies where death was not an end but a transition to another form of existence.

Let's delve into this fascinating topic by first understanding how life might manifest itself across multiple realities.

Life in Infinite Realities

The multiverse theory suggests infinite possibilities for how life could evolve or exist. For instance, consider a universe where evolution took a slightly different turn than it did on Earth. Perhaps dinosaurs never went extinct there and evolved into intelligent beings ruling their world? Or imagine a reality where silicon-based life forms thrive instead of carbon-based ones like us?

Sam once stumbled upon such an alien world during his travels—a planet teeming with crystalline creatures whose bodies refracted sunlight into dazzling displays of color—an entire civilization built from living gemstones!

Such examples challenge our conventional understanding about what constitutes 'life.' They force us to expand our definitions beyond earthly norms—beyond carbon chemistry or even physical bodies— to include entities that may be purely energy-based or exist within virtual realms.

Death Across Dimensions

Now let's explore the other side of this existential coin—death—in these myriad realities.

In one parallel universe Sam visited, people didn't die—they merely shifted their consciousness to younger clones created at birth. This society viewed death not as an inevitable biological eventuality but as something preventable—a disease they had cured long ago!

On another sojourn through space-time continuum, Sam encountered a civilization that had transcended physical forms altogether. Their consciousness existed within a vast digital network, and 'death' was merely the shutting down of an outdated data node.

In yet another reality, death was not feared but celebrated as a passage to an alternate dimension—a spiritual realm where beings continued their journey in different forms.

Exercises for Readers

* * *

1. Imagine you're Sam Doe, exploring these infinite realities. How would encountering such diverse perspectives on life and death affect your own views? Write down your thoughts.

2. Create your own parallel universe with unique concepts of life and death—be as creative or scientifically plausible as you want!

The Hero's Journey Continues

As our hero traverses through these dimensions, he learns more than just the scientific wonders of the multiverse—he experiences firsthand how fluid concepts like life and death can be when viewed from different lenses.

These encounters profoundly change him—they make him question his beliefs about mortality, existence, even purpose. They teach him humility before the grandeur of creation while also empowering him with knowledge beyond any single world's reach.

Sam realizes that in this boundless cosmos teeming with endless possibilities, every end is but a new beginning—an echo reverberating across infinite realities—and perhaps true immortality lies not in evading death but embracing it as part of this cosmic dance.

This chapter serves to remind us all: As we navigate through our lives —one among countless others—we must remember that each moment is precious; each decision creates ripples across multiple universes; every life matters—even if it exists within one tiny speck floating in an endless sea called Multiverse.

LOVE ACROSS DIMENSIONS - INTERDIMENSIONAL RELATIONSHIPS

Chapter 38: Love Across Dimensions - Interdimensional Relationships

In the vast expanse of the multiverse, where infinite realities coexist, one might wonder about the nature of relationships and love. How does love manifest in a reality different from ours? Can it transcend dimensions? The concept may seem far-fetched or straight out of a science fiction novel, but as we delve deeper into understanding parallel universes, these questions become increasingly relevant.

Let's start by considering our hero's journey through multiple worlds. He has encountered various versions of people he knows – friends, family members, even himself. But what happens when he meets someone special in another universe? Someone who stirs his heart and makes him feel emotions that are both familiar yet strange?

Imagine this scenario: Our hero finds himself in an alternate reality where everything is similar to his own world except for one significant difference – here exists a woman who captures his heart instantly. She is not present in his original universe but here she stands before him as real as anyone else.

This encounter leads us to ponder on interdimensional relationships. Would feelings remain consistent across different realities? If you fall in love with someone in one universe, would those feelings persist if you met their counterpart elsewhere?

To explore this further let's consider an example from pop culture; specifically "The One," a Jet Li movie which presents an interesting take on inter-dimensional romance. In the film, Jet Li plays two characters - Gabe Law and Yulaw - each living separate lives across different dimensions until circumstances bring them together.

Gabe Law lives a peaceful life with his wife T.K., while Yulaw is hell-bent on eliminating all other versions of himself across multiple universes to gain absolute power. When Yulaw arrives at Gabe's dimension intending to kill him too, he encounters T.K., leading to unexpected emotional turmoil within him despite having never met her version anywhere else.

This scenario raises intriguing questions about the nature of love and relationships across dimensions. Is it possible that Yulaw, despite his ruthless demeanor, could harbor feelings for T.K., a woman he has never met before? Could there be an inherent connection between souls that transcends physical realities?

The concept of soulmates often suggests that two people are meant to be together regardless of their circumstances. If we apply this idea to our multiverse theory, it's fascinating to consider whether these connections could span across different universes.

However, interdimensional relationships also present unique challenges. For instance, how would one deal with the knowledge that their loved one exists in another universe but not theirs? Or what if they meet a version of their partner who is significantly different from the person they fell in love with?

As our hero navigates through these complex emotions and experiences, he learns valuable lessons about love and its boundless nature. He realizes that while each universe may have its own set rules and laws governing life, certain things like love remain constant.

In conclusion, exploring interdimensional relationships allows us to broaden our understanding of human emotions beyond the confines of our known reality. It encourages us to question conventional notions about love and opens up new avenues for thought-provoking discussions on human connections in a multiverse setting.

As you continue your journey through this book (and perhaps even parallel universes), ponder upon these ideas: Can true love transcend space-time boundaries? Are some bonds destined irrespective of which universe they exist within? And most importantly - can understanding such concepts change how we perceive and experience love in our own world?

THE POWER OF CHOICE IN A DETERMINISTIC UNIVERSE

Chapter 39: The Power of Choice in a Deterministic Universe

* * *

In the previous chapters, we've explored the vastness and complexity of multiverses. We've journeyed through dimensions, encountered alien civilizations, and even faced battles across realities. Now let's delve into one of the most profound philosophical questions that arise from our understanding of multiverses - do we have free will or are our lives predetermined by cosmic laws?

Imagine for a moment that you're standing at an intersection in your life where two roads diverge. One path leads to a future filled with joy and success while the other is fraught with challenges and hardships. In a deterministic universe, it would seem that your choice has already been made for you by some unseen force or law.

But what if I told you that within this deterministic universe lies another layer – one where every possible outcome exists simultaneously? This is not just theoretical musing but grounded in quantum mechanics' Many-Worlds Interpretation (MWI). According to MWI, each decision creates a split in reality where all outcomes occur.

Let's return to our main character's journey as he navigates through these parallel universes. He finds himself facing an interdimensional villain threatening his home world's existence. Our hero must make choices - which strategy to employ, who to trust, when to fight or retreat – knowing full well each decision branches off into different realities.

Consider this scenario like playing chess against yourself on multiple boards simultaneously; every move spawns new games with varying results. But unlike chess pieces bound by rules on movement and capture, humans possess consciousness allowing us insight into potential consequences before making decisions.

This brings us back to our question about free will versus determinism within multiverse context. If every choice creates new universes accommodating all possibilities then aren't we exercising free will? Yet paradoxically isn't everything still predestined since those universes already exist?

To reconcile this, we must redefine our understanding of free will. It is not merely the ability to choose between A or B but rather a conscious navigation through the multiverse landscape based on our values, desires, and wisdom.

Let's take an example from pop culture - Marvel's "Doctor Strange". In the movie, Doctor Strange uses his time stone to view millions of possible futures against their enemy Thanos. Despite knowing only one scenario leads to victory he still fights for it. Isn't this a perfect illustration of choice within determinism?

To further explore this concept let's do an exercise:

1) Think about a major decision you've made recently.
    2) Now imagine all possible outcomes that could have resulted from different choices.
    3) Reflect on why you chose your particular path over others.

This exercise helps us realize that even in a deterministic universe with infinite possibilities we are not mere passive observers but active participants shaping our reality.

In conclusion, while multiverses might seem dauntingly complex they also empower us by highlighting the importance and power of choice amidst apparent determinism. As physicist Brian Greene said "The future is fixed in the sense that it exists as surely as does the present", yet how we arrive there remains up to us.

Our hero understands this now more than ever as he stands ready for his final battle – each move calculated yet uncertain; every strategy branching off into new realities; aware that despite cosmic laws dictating existence across universes ultimately his choices matter...and so do yours!

* * *

THE ROLE OF CHANCE AND PROBABILITY IN THE MULTIVERSE

Chapter 40: The Role of Chance and Probability in the Multiverse

In our everyday lives, we often encounter situations where chance and probability play a significant role. From flipping a coin to predicting weather patterns, these concepts are deeply ingrained in our understanding of the world around us. But what happens when we extend these ideas to the realm of parallel universes? How does chance factor into an infinite number of realities?

Let's begin by revisiting our hero's journey through multiple dimensions. By now, he has navigated countless worlds, each with its unique set of rules and circumstances. He has seen worlds where gravity is reversed, time flows backward, or even places where dinosaurs still roam freely.

But one question keeps nagging at him - why? Why do these different realities exist as they do? Is it all just random chaos or is there some underlying order guiding their formation?

To answer this question let's delve into quantum mechanics – specifically the concept known as superposition. This principle suggests that until observed or measured, particles exist in all possible states simultaneously - like Schrödinger's famous cat being both alive and dead at once.

Imagine flipping a coin but not looking at it immediately after you flip it; according to quantum theory (and assuming for simplicity that your coin can't land on its edge), your coin is both heads AND tails until you look at it! Now apply this idea across an infinite multiverse - every possible outcome exists somewhere out there!

This brings us back to our hero who finds himself standing before two identical doors leading to different universes—one where his greatest dreams come true and another filled with his worst nightmares.

* * *

The choice seems obvious—but remember—this isn't about making 'the right' decision—it's about understanding how choices work within the framework of infinity itself! In essence—every choice he could make—he HAS made—in some universe somewhere!

Now imagine if instead of two doors, there were infinite doors—each leading to a different universe. The probability of our hero choosing any particular door becomes infinitesimally small yet remains possible.

This is where the concept of chance comes into play in the multiverse. Each decision, each random event that occurs creates a new branch—a new world—in this vast cosmic tree. It's like rolling an infinite-sided dice—the outcome is unpredictable and every roll generates a whole set of parallel realities!

But what does this mean for our hero? And more importantly—for us?

It means that while we may not have control over the probabilities—we do have control over our choices! We can't predict which universe we'll find ourselves in next—but we can choose how to act within it.

In essence, understanding chance and probability in the multiverse empowers us rather than diminishes us—it shows us that even amidst chaos—there's room for choice—and therefore—room for hope!

As you ponder upon these ideas, consider this: What if right now—you're standing before an infinity of doors? Which one will you choose?

Exercise: Reflect on your life decisions so far - big or small. Now imagine alternate versions of yourself who made different choices at those critical junctures. How would their lives be different from yours? This exercise isn't meant to induce regret but rather broaden your perspective about life's possibilities.

MIND-BENDING PARADOXES AND PUZZLES

Chapter 41: Mind-Bending Paradoxes and Puzzles

In our journey through the multiverse, we've encountered many strange phenomena. We've seen worlds where time runs backward, universes with different physical laws, and even realities where our most cherished beliefs are turned upside down. But perhaps nothing is as perplexing or mind-bending as the paradoxes and puzzles that arise when contemplating the nature of parallel universes.

Let's start by considering one of the most famous paradoxes in all of science fiction – The Grandfather Paradox. This thought experiment asks what would happen if you traveled back in time and killed your own grandfather before he had a chance to father your parent. If your grandfather died before having children, then you could never have been born... but if you were never born, how could you go back in time to kill him?

This is an example of a temporal paradox - a situation where cause-and-effect become tangled up in such a way that they seem to contradict each other. In some interpretations of quantum mechanics (such as the Many-Worlds Interpretation), these kinds of paradoxes are resolved by suggesting that traveling back in time actually creates an entirely new "branch" universe separate from ours.

Our hero faced similar conundrums during his adventures across dimensions. Remember when he found himself on Earth-27B? There, he met another version of himself who had made drastically different life choices due to minor differences earlier in their shared timeline.

The existence of this doppelgänger raised puzzling questions about identity and individuality: Who was 'real'? Were they both equally real because they existed within their respective universes? Or was one more 'authentic' than the other? These existential riddles forced our hero not only to question his understanding but also his sense-of-self.

Another fascinating puzzle arises when considering communication between parallel worlds - let's call it The Interdimensional Message Paradox. Suppose you could send a message to another universe, and they could respond. You might be tempted to ask them for information about the future (since some universes may be ahead of ours in time), but this raises all sorts of thorny issues.

For instance, if they tell us about a disaster that's going to happen and we prevent it, then from their perspective, it would seem like our actions invalidated their past knowledge - another paradox! This is similar to the "Bootstrap Paradox" or causal loop situation where information or objects can exist without having been created.

As we delve deeper into these mind-bending puzzles and paradoxes, let's remember what Albert Einstein once said: "The most incomprehensible thing about the universe is that it is comprehensible." Even when faced with seemingly impossible conundrums, human curiosity and ingenuity find ways to understand and make sense of our reality – or should I say realities?

Now here's an exercise for you: Think about your own life as one path among many in a multiverse. How does considering other possible versions of your life help illuminate who you are now? What choices have defined your current 'universe'? And how might different decisions have led you down entirely different paths?

In the next chapter, we'll explore spirituality and religion across different realities. But before then, take some time to ponder upon these paradoxes because sometimes questions are more important than answers.

SPIRITUALITY AND RELIGION IN A BOUNDLESS COSMOS

Chapter 42: Spirituality and Religion in a Boundless Cosmos

In the vast expanse of the multiverse, where countless realities coexist, each with its unique set of physical laws and life forms, one might wonder about the place of spirituality and religion. How do these age-old human constructs fit into this boundless cosmos? What happens to our beliefs when we are faced with an infinite number of possibilities?

Our hero had been grappling with similar questions as he journeyed through different dimensions. He had seen worlds where religions were non-existent, replaced by a profound understanding of cosmic unity. In others, faith was deeply ingrained in every aspect of life.

Let's take a step back for a moment and consider what religion means to us here on Earth. For many people around the world, their religious beliefs provide them with moral guidance and spiritual comfort. They offer answers to existential questions such as why we exist or what happens after death.

Now imagine being confronted with an infinite array of universes - some eerily similar to ours but slightly off-kilter; others so radically different they defy comprehension. Would our earthly religions still hold up under such circumstances? Or would they crumble under the weighty realization that we are but one tiny speck within an endless cosmic sea?

Consider Christianity's belief in resurrection or Hinduism's concept reincarnation – how would these doctrines fare in parallel universes? Could there be multiple versions of you living out various lives simultaneously across different dimensions according to Eastern philosophies like Buddhism or Jainism?

The hero encountered civilizations that had developed entirely new spiritual systems based on their understanding of the multiverse— religions centered around quantum mechanics or string theory instead traditional deities.

* * *

But it wasn't just established religions that came under scrutiny during his travels; even atheistic perspectives were challenged by encounters with advanced alien species who possessed abilities akin god-like powers from human perspective.

As he delved deeper into these questions, our hero realized that the multiverse concept doesn't necessarily negate spirituality or religion. Instead, it expands them into new territories.

For instance, in a universe where time is non-linear, religions that believe in reincarnation could see this as validation of their beliefs. In another universe where multiple versions of individuals exist simultaneously, we might find support for spiritual concepts like soul families or group karma.

The hero also discovered societies which had integrated science and spirituality seamlessly. They didn't see any conflict between believing in a higher power and understanding the intricacies of quantum physics or cosmic inflation. For them, scientific discoveries were just another way to comprehend the divine mystery.

In essence, what he learned was that spirituality and religion are not static constructs but evolve with our understanding of reality itself. As we continue to unravel the mysteries of the cosmos—be it through telescopes peering into distant galaxies or theories postulating parallel universes—we may need to reevaluate and expand our spiritual perspectives accordingly.

To help you ponder over these profound issues here's an exercise: Imagine yourself living in a different universe with its unique set of physical laws. How would your belief system adapt? Would you cling onto your current faith or develop new ways to understand existence?

As Albert Einstein once said: "Science without religion is lame; religion without science is blind." Perhaps as we venture further into exploring multiverses—a realm where science fiction meets reality—

we will find ways to harmonize our thirst for empirical knowledge with our innate longing for spiritual connection.

AESTHETICS, ART, AND BEAUTY ACROSS DIFFERENT REALITIES

Chapter 43: Aesthetics, Art, and Beauty across Different Realities

Art is a universal language. It transcends boundaries of time, space, culture and even realities. But what happens when we consider art in the context of multiple universes? How does aesthetics change when viewed through the lens of parallel worlds?

Our hero had seen many things on his journey through the multiverse - alien civilizations with technology far beyond our comprehension, worlds where history took a different turn leading to alternate present scenarios. However, nothing prepared him for the sheer diversity he would encounter in terms of aesthetics and beauty.

In one universe he found himself standing before an enormous canvas that stretched as far as his eyes could see. The painting was alive – literally! Each brush stroke pulsed with life energy; colors morphed and shifted creating an ever-changing tableau that reflected its viewers' emotions back at them.

Another reality presented music so complex it seemed to have extra dimensions to it - not just length or breadth but depth too. This multidimensional symphony resonated within him at frequencies he didn't know existed.

These experiences made him question - What is beauty? Is there a universal standard for aesthetic appeal or does it vary from universe to universe?

Consider this: In our world we often equate symmetry with beauty due to evolutionary reasons – symmetrical faces are considered more attractive because they indicate good health and genetic fitness. But what if you were in a universe where survival depended on asymmetry? Would their standards of beauty reflect that?

Or think about color perception which depends on how light interacts with our eyes. Some animals can see ultraviolet light while others perceive more shades of green than us humans do. Now imagine being in a world where inhabitants perceived an entirely different spectrum of light! Their art would be incomprehensible to us yet breathtakingly beautiful by their standards.

This brings us back full circle: Beauty truly lies in the eye (or sensory organ) of the beholder. It is subjective, relative and ever-changing.

Our hero's journey through different realities taught him to appreciate this diversity. He learned that art isn't just about creating something beautiful; it's a way of seeing the world (or worlds). Each universe had its own unique aesthetic language which reflected their history, culture and even physics!

So next time you look at a piece of art, try to see beyond your immediate perception. Consider what it might mean in another context or reality. You never know - you might catch a glimpse into another universe!

Exercise for readers: Take an artwork you love – could be anything from Mona Lisa to Starry Night – and reimagine it in the context of a parallel universe with different physical laws or cultural norms. How would these changes affect your interpretation? Share your thoughts on social media using #MultiverseArtChallenge.

Remember: Art is not bound by our limited understanding of reality but rather expands it by offering glimpses into infinite possibilities.

COPING WITH EXISTENTIAL CRISIS

Chapter 44: Coping with Existential Crisis

In the vast expanse of the multiverse, our hero found himself grappling with a profound existential crisis. The sheer magnitude and complexity of infinite realities had begun to weigh heavily on his mind. He was but a speck in an endless sea of possibilities, each one as real and tangible as his own existence.

The concept of the multiverse can indeed be daunting. It challenges our very perception of reality and forces us to confront questions that we may not have answers for. What is my purpose? Do I matter? Is there any point to all this?

Our hero wasn't alone in these feelings; many great minds throughout history have grappled with similar existential crises when faced with the enormity of our universe - let alone multiple universes! Albert Einstein once said, "One may say 'the eternal mystery of the world is its comprehensibility.'" This statement encapsulates perfectly how confronting such grand concepts can lead us down a path towards existential questioning.

So how does one cope when faced with an existential crisis triggered by contemplating parallel worlds?

Firstly, it's essential to understand that feeling overwhelmed or insignificant in light of such revelations isn't necessarily negative. These feelings are natural responses to new information that dramatically shifts our perspective about ourselves and our place within reality (or realities).

Secondly, remember that while we might be small parts within an infinitely larger whole, this doesn't diminish our value or worth. Each universe within the multiverse has its unique properties and events – including ours – making every single entity within it inherently valuable due to their uniqueness.

Consider Vincent Van Gogh's painting "Starry Night." In it, he captures countless stars twinkling against a dark sky—a representation perhaps not too dissimilar from imagining numerous universes scattered across dimensions. Yet amongst those stars lies Earth—small yet significant because it's home.

Similarly, you're significant because you're the protagonist of your universe. Your thoughts, actions, and experiences matter because they contribute to the uniqueness of your reality.

Our hero realized this as he journeyed through different dimensions. He saw that while his existence might be one among many, it was still unique—filled with its own triumphs and tribulations. This realization helped him find meaning amidst chaos.

Thirdly, use these existential questions as a catalyst for personal growth. Instead of succumbing to despair or nihilism, let them inspire curiosity and wonder about the nature of existence itself.

As Carl Sagan once said: "Somewhere, something incredible is waiting to be known." The multiverse theory opens up infinite possibilities for discovery and exploration—not just in terms of physical realities but also our understanding of consciousness, identity, purpose, and life's interconnectedness.

In conclusion: when faced with an existential crisis brought on by contemplating parallel worlds or any grand concept that challenges our perception—embrace it! Use it as a tool for self-reflection and growth; remember your inherent worth within this vast cosmos; stay curious; keep exploring—and like our hero—you too will navigate successfully through the labyrinthine complexities posed by living in a multiverse.

FACING FEAR – CONFRONTING UNKNOWN ENTITIES

Chapter 45: Facing Fear – Confronting Unknown Entities

* * *

Fear is a primal instinct, an evolutionary response to potential danger. It's what kept our ancestors alive in the face of predators and other threats. But fear can also be paralyzing, especially when it comes to confronting the unknown.

In our journey through the multiverse, we've encountered many strange and unfamiliar entities. Some are benign or indifferent; others pose significant challenges that test our courage and resolve. Our hero has faced these fears head-on, demonstrating remarkable bravery in his quest for knowledge and understanding.

Let us take a moment to reflect on one such encounter...

Our hero found himself standing at the edge of a swirling vortex - a portal leading into another universe entirely different from ours. As he peered into its depths, he felt an overwhelming sense of dread creeping over him. The world beyond was filled with shadowy figures moving about in ways that defied logic and reason.

These were not just physical beings but manifestations of concepts alien to human comprehension - abstract entities embodying paradoxes, impossibilities, contradictions... They were terrifying precisely because they challenged everything our hero thought he knew about reality.

But instead of succumbing to fear, our hero chose confrontation over retreat. He stepped forward bravely into this new realm armed with curiosity as his shield and determination as his sword.

His first encounters were disorientating—like trying to communicate without common language or shared context—but gradually he began learning their ways by observing patterns in their behavior and interactions.

He discovered that some entities represented mathematical principles while others embodied philosophical concepts or natural phenomena from their home universe—a fascinating blend of science fiction meets philosophy!

This experience taught him valuable lessons about facing fears:

1) **Acknowledge your fear**: Recognizing your emotions is crucial before you can address them effectively.

2) **Understand its source**: What exactly are you afraid of? Is it rational?

3) **Prepare yourself**: Equip yourself with knowledge and tools to face the challenge.

4) **Take action**: Fear diminishes when confronted directly.

Now, let's put these lessons into practice. Imagine you're about to step into a parallel universe filled with unknown entities.

Exercise 1: Acknowledge your fear

Write down what you feel as you stand before this portal. Is it excitement? Anxiety? A mix of both?

Exercise 2: Understand its source

Identify what exactly is causing your fear. Is it the uncertainty of what lies ahead? The potential dangers that might lurk in this new world?

Exercise 3: Prepare yourself

List down all the skills, knowledge, and resources you have that could help you navigate through this unfamiliar territory.

Exercise 4: Take action

Visualize yourself stepping forward bravely into the portal, ready to confront whatever comes your way.

Remember Sam's journey—his courage didn't mean he was never afraid; instead, he chose not to let his fears dictate his actions. He faced them head-on and emerged stronger each time.

As we continue exploring multiverses teeming with unknown entities, remember these lessons from our hero's journey - they are not just applicable for interdimensional travel but also for confronting fears in our everyday lives!

In conclusion:

"Fear is a reaction; courage is a decision." - Winston Churchill

So decide wisely!

NURTURING HOPE AMIDST CHAOS

Chapter 46: Nurturing Hope amidst Chaos

In the vast expanse of infinite realities, our hero found himself in a world that was on the brink of collapse. The once vibrant cityscape had turned into ruins and despair hung heavy in the air. It was a stark contrast to his home universe where peace and prosperity reigned supreme.

The concept of hope might seem outlandish when faced with such dire circumstances. Yet, it is precisely during these times that nurturing hope becomes an act of defiance against chaos and destruction.

Hope is not just wishful thinking; it's a form of resilience, an affirmation that even in the face of overwhelming odds, we can still envision a better future. As Emily Dickinson eloquently put it, "Hope is the thing with feathers That perches in the soul And sings the tune without words And never stops at all."

Our hero realized this as he navigated through this dystopian reality. He saw people clinging onto fragments of their past lives while trying to build something new from those remnants - like phoenixes rising from ashes.

Take for instance Rosa Parks who refused to give up her seat on a bus or Mahatma Gandhi who led India towards independence through non-violent resistance – they were ordinary individuals who dared to dream big despite being surrounded by turmoil.

Similarly, our protagonist decided not only to survive but also inspire change within this parallel universe. He started small by helping rebuild homes using debris and salvaged materials – physical manifestations symbolizing renewed hopes for many inhabitants there.

This chapter invites you too as readers to reflect upon your own life situations - Are there instances where you've felt overwhelmed by chaos? How did you keep your hopes alive?

Here are some exercises:

1) Write down three challenging situations you've encountered recently.

   2) For each situation, list potential positive outcomes or opportunities for growth.

   3) Reflect on how maintaining hope could influence your perspective towards these challenges.

Remember, nurturing hope amidst chaos is not about denying reality or avoiding difficulties. It's about acknowledging the situation and choosing to believe that better days are ahead.

Our hero's journey in this chaotic world was far from over. But he had learned an invaluable lesson – Hope, like a beacon of light, can guide us through even the darkest of times. As we traverse our own multiverses filled with trials and tribulations, let us remember to carry this torch of hope along with us.

In the words of Martin Luther King Jr., "We must accept finite disappointment but never lose infinite hope." This chapter serves as a testament to that spirit - A reminder that no matter how tumultuous our journey across parallel universes might be, there's always room for nurturing hope amidst chaos.

BRIDGING SCIENCE FICTION WITH REALITY

Chapter 47: Bridging Science Fiction with Reality

In the realm of science fiction, the impossible becomes possible. Alien civilizations flourish, time travel is a weekend pastime, and parallel universes are just a wormhole away. But how much of this fantastical world can be bridged with our reality? How close are we to turning these imaginative concepts into tangible experiences?

Our hero has journeyed through countless realities, faced unimaginable challenges and returned home bearing wisdom from across dimensions. He's lived what most would consider pure fantasy. Yet his experiences aren't as far removed from our own potential future as one might think.

Let's start by examining some examples where science fiction has already become reality.

Consider Jules Verne's "20,000 Leagues Under the Sea," published in 1870 which predicted submarine warfare - a concept that materialized during World War I. Or Arthur C Clarke's geostationary satellites idea in his work "Wireless World," which became an integral part of modern communication systems.

Closer to our topic at hand is the concept of parallel universes itself – once purely speculative and now considered seriously by physicists worldwide thanks to theories like quantum mechanics and string theory.

The bridge between science fiction and reality lies within human imagination coupled with scientific curiosity. It starts as an abstract thought or creative storytelling but ends up inspiring real-world research leading towards actualization.

* * *

Take for instance teleportation - popularized by Star Trek's famous phrase "Beam me up Scotty". While we're not yet able to transport humans instantaneously through space (and certainly not without scrambling them into atoms), scientists have successfully teleported information using principles of quantum entanglement - proving that even seemingly outlandish ideas may hold grains of truth waiting for discovery.

Now let's turn back to our hero who stands at this very bridge between sci-fi dreams and scientific possibility. His adventures across multiple realities were thrilling indeed but could they ever become part of our reality?

The answer lies in the progress we make in quantum physics, cosmology and technology. While we're still far from physically traversing parallel universes, research into quantum computers is paving the way for us to simulate complex multiverse scenarios.

Moreover, advancements in telescopes and satellite technologies are allowing us to peer deeper into space than ever before - who knows what secrets about our universe (or perhaps others) await discovery?

As a practical exercise, let's engage with science fiction more actively. Read a sci-fi novel or watch a movie that explores concepts of parallel universes. As you do so, question how these ideas might be realized scientifically. What theories would support them? What technological advances would be needed? This will not only deepen your understanding but also stimulate your imagination towards future possibilities.

Our hero's journey may seem fantastical now but remember – so did submarines and satellites once upon a time! The bridge between science fiction and reality is there waiting for daring dreamers like him (and you!) to cross it.

TOWARDS AN INCLUSIVE COSMOLOGY

Chapter 48: Towards an Inclusive Cosmology

Our journey through the multiverse has been nothing short of extraordinary. We've traversed dimensions, encountered alien civilizations, and even grappled with existential crises that have stretched our understanding of reality to its limits. But as we stand on the precipice of uncharted territories, it's time to take a step back and reflect on what this all means for us - not just as individuals or even as a species, but in terms of our place within the grand cosmic tapestry.

Inclusivity is often thought about in social contexts – ensuring equal opportunities regardless of race, gender or socioeconomic status. However, when applied to cosmology – the study of the universe's origins and evolution – inclusivity takes on a whole new meaning.

The concept here is simple yet profound: every single entity in existence - from galaxies down to subatomic particles - plays an integral part in shaping reality. Our hero learned this firsthand during his adventures across parallel worlds; each decision he made had ripple effects throughout multiple realities.

Consider for instance how dark matter and dark energy were initially perceived as anomalies because they didn't fit into existing cosmological models. Instead of dismissing them outright though, scientists expanded their theories to accommodate these mysterious phenomena thereby making cosmology more inclusive.

This approach isn't limited to scientific endeavors alone; pop culture too has played a significant role by popularizing concepts like parallel universes thus democratizing access to complex ideas once confined within academic circles.

Now let's turn towards our main character who started off as an ordinary individual unaware about other realities co-existing alongside ours. His transformation from being merely curious about alternate universes into becoming a master navigator through different dimensions serves as an allegory for humanity's potential growth if we embrace inclusive cosmology fully.

Imagine if everyone understood that their actions could potentially affect countless lives across infinite realities? Would people be more mindful before taking decisions? Could this awareness foster a sense of unity and shared responsibility towards preserving not just our own world but other universes too?

Inclusive cosmology also challenges us to redefine what we consider as 'alien'. Our hero encountered beings from different dimensions, each with their unique cultures and ways of life. Instead of reacting with fear or hostility, he learned to appreciate the diversity that exists within the multiverse.

This is an important lesson for humanity as well; if we ever come across extraterrestrial life forms, it's crucial that we approach them with open minds rather than preconceived notions based on our limited understanding.

As you ponder over these questions, here's an exercise: imagine yourself in a parallel universe where your life took a completely different turn due to one decision made differently. How would you feel about meeting your alternate self? What could you learn from each other's experiences?

Our journey through the multiverse has been enlightening indeed but remember - it doesn't end here. As long as there are unexplored realities out there waiting to be discovered, our quest for knowledge continues... because after all, isn't that what inclusive cosmology is all about – acknowledging every piece of the puzzle no matter how small or seemingly insignificant?

So let's keep pushing boundaries and venturing into unknown territories because who knows - maybe somewhere out there in another reality lies answers to some of our most pressing questions...or perhaps even solutions to problems plaguing our world today.

PIONEERING INTO UNCHARTED TERRITORIES

Chapter 49: Pioneering into Uncharted Territories

As we delve deeper into the mysteries of the multiverse, we find ourselves standing on the precipice of uncharted territories. It's akin to being an explorer in a new world, where every step forward is a step into unknown realms. But isn't that what makes this journey so exhilarating?

Our hero has come a long way from his humble beginnings as an ordinary individual to becoming a seasoned traveler across dimensions. He has faced challenges and overcome obstacles that would have seemed insurmountable at first glance.

Now he stands ready for his next big adventure - pioneering into unexplored regions of the multiverse.

The concept of exploration is not new to humanity. We've always been explorers by nature, driven by curiosity and desire to understand our surroundings better. From early humans venturing out of Africa to modern astronauts setting foot on the moon, history is replete with examples of brave individuals who dared to venture beyond known boundaries.

Take Christopher Columbus for instance; despite facing numerous setbacks and uncertainties, he embarked on an ambitious voyage across the Atlantic Ocean in search for Asia but ended up discovering America instead. Or consider Neil Armstrong; amidst great risks and potential dangers, he took "one small step" onto lunar surface thereby marking mankind's giant leap towards space exploration.

Similarly, exploring parallel universes presents its own set of unique challenges and opportunities. The stakes are high but so are potential rewards.

Imagine finding worlds where laws of physics operate differently or encountering civilizations far advanced than ours! What if there exist realities where extinct species still thrive or historical events had different outcomes? The possibilities seem endless!

However, such exploration requires more than just courage and curiosity; it demands innovation too! Just like how invention of compass facilitated sea voyages or development rocket technology enabled space travel, navigating through multiple realities necessitates breakthroughs in quantum mechanics & string theory among others.

Moreover it's not just about scientific advancements but also ethical considerations. How do we interact with beings from other dimensions? What are our responsibilities towards them and their worlds?

Our hero, having experienced numerous realities, understands these complexities well. He knows that every action can have far-reaching consequences across multiple universes due to butterfly effect.

Yet he is undeterred because he believes in the power of knowledge & wisdom gained through his travels. He has seen firsthand how understanding different perspectives can lead to growth and progress.

As he prepares for this new journey, he reflects upon a quote by T.S Eliot: "We shall not cease from exploration, and the end of all our exploring will be to arrive where we started and know the place for the first time."

Indeed, as we pioneer into uncharted territories of multiverse, perhaps what we'll discover most is ourselves - our potentialities & limitations; our fears & hopes; our pasts & futures.

* * *

So let's embark on this exciting adventure together! Let's dare to dream big! Let's pioneer into uncharted territories!

Exercise:

1) Imagine you're an explorer venturing into a parallel universe for the first time. Write down your feelings.

2) Think about some ethical dilemmas you might face during such explorations.

3) Reflect upon how encountering different realities could influence your worldview.

4) Consider ways in which such multidimensional travel could impact human civilization as whole.

5) Ponder over technological innovations required for navigating through multiverses.

DREAMERS & VISIONARIES – SHAPING FUTURE NARRATIVES

Chapter 50: Dreamers & Visionaries – Shaping Future Narratives

In the grand tapestry of existence, there are those who follow the threads already woven and those who dare to weave new ones. These individuals, dreamers and visionaries as we call them, have a unique ability to shape future narratives. They see beyond what is present and imagine what could be.

Our hero had become one such visionary after his adventures across multiple universes. He had seen realities that defied comprehension, encountered beings of unimaginable intelligence, witnessed histories diverging from our own in fascinating ways. This chapter explores how these experiences shaped him into a visionary capable of influencing not just his original universe but countless others he'd come to know.

The concept here is simple yet profound - shaping future narratives through imagination and innovation. It's about using our creative faculties to envision different futures and then working towards making them a reality.

Consider for instance the story of Elon Musk, an entrepreneur whose visions have been reshaping industries ranging from electric vehicles with Tesla Motors to space travel with SpaceX. His ambitious goal of colonizing Mars may seem like science fiction today but so did many technological advancements before they were realized.

Similarly, authors like H.G Wells or Jules Verne imagined technologies in their stories long before they became real-world inventions; submarines, time machines or moon landings were all once figments of imaginative minds before becoming tangible realities.

Drawing parallels between these visionaries' work and our hero's journey can provide valuable insights on how one might influence future narratives across multiple dimensions.

Firstly it's important to understand that being a visionary isn't merely about having big ideas; it also involves taking action towards realizing them despite challenges or setbacks - much like our hero did when navigating through parallel worlds.

Secondly visionaries often draw upon their diverse experiences for inspiration - just as our protagonist used knowledge gained from various universes to solve problems within each new reality he encountered.

Finally, visionaries understand the power of storytelling. They know that narratives can inspire change, create empathy and drive innovation. Our hero too realized this as he shared tales from his adventures across dimensions - each story serving to broaden perspectives and challenge preconceived notions within his home universe.

Now let's consider some exercises for you, dear reader:

1. Imagine a parallel world: What does it look like? Who lives there?

What are their customs or technologies?

2. Now think about how experiences from this imagined world could be used to solve a problem in our current reality.

3. Finally write a short narrative set in your parallel world – remember the power of storytelling!

In conclusion, being a visionary is not exclusive to those with extraordinary intelligence or resources; it's about daring to dream beyond what exists and taking steps towards making those dreams come true - just like our hero did on his journey through the multiverse.

As we move forward into uncharted territories of knowledge and discovery, may we all become dreamers & visionaries shaping future narratives across multiple realities!

LAYING FOUNDATIONS FOR INTERDIMENSIONAL TRAVEL

Chapter 51: Laying Foundations for Interdimensional Travel

In the previous chapters, we have explored the concept of multiverse and parallel universes. We've journeyed through different dimensions, encountered alien civilizations, and even navigated through cross-dimensional crises. Now it's time to tackle a question that has been lingering in our minds since the beginning - How can we travel between these multiple realities?

Our hero, who started as an ordinary individual but now stands as a king amongst many worlds, has already demonstrated this possibility. But how can we lay down foundations for interdimensional travel on a larger scale? Let's delve into this fascinating topic.

Firstly, let us understand that interdimensional travel is not just about physical movement from one place to another; it involves transitioning from one reality to another. This requires understanding and manipulating the very fabric of space-time itself.

Theoretical physicists propose several methods for possible interdimensional travel. One such method is through wormholes – hypothetical shortcuts through space-time predicted by Einstein's theory of general relativity. Imagine folding a piece of paper (representing space-time) so two points touch each other – that connection point would be akin to a wormhole.

However, creating stable wormholes presents significant challenges due to their tendency towards instability and collapse under normal conditions. The introduction of exotic matter with negative energy density could potentially stabilize them but obtaining or creating such matter remains purely theoretical at present.

Another proposed method involves harnessing quantum entanglement – when particles become interconnected regardless of distance separating them in space or time dimensionally speaking they are always 'next door' neighbors! Could this principle be extended beyond microscopic particles?

Let's take inspiration from our hero's journey here: he didn't start off knowing how to navigate across dimensions; instead he learned along his path encountering various trials tribulations which shaped him into master navigator today Similarly humanity might need undergo its own collective learning process before mastering interdimensional travel.

This brings us to the importance of education and research. We need to invest in scientific exploration, encourage curiosity about the universe, and foster a culture that values knowledge and discovery. The more we understand about our universe (or rather multiverse), the closer we get to making interdimensional travel a reality.

Moreover, it's not just about technological advancements but also psychological readiness. Interacting with different realities could be disorienting or even terrifying. Remember when our hero first encountered an alternate universe? It was overwhelming! Therefore, mental preparedness is crucial for future interdimensional travelers.

In addition, ethical considerations must be addressed: What are consequences of interfering with another reality? How can we ensure respect for life forms in other dimensions?

As we ponder these questions let's remember words Albert Einstein "Imagination everything preview life's coming attractions" Perhaps key unlocking secrets interdimensional travel lies within power human imagination combined rigorous scientific inquiry

So as lay foundations this monumental endeavor let us embrace spirit adventure courageously step into unknown After all isn't what makes journey worthwhile?

Now dear reader I invite you imagine yourself as an explorer new worlds What would your first steps look like? How would you prepare for such a journey? And most importantly what kind world(s) do hope discover?

EVOLVING AS AN INTERGALACTIC SPECIES

Chapter 52: Evolving as an Intergalactic Species

As we delve deeper into the mysteries of the multiverse, one question that persistently arises is how humanity might evolve to become a truly intergalactic species. This chapter will explore this concept in depth, examining both the scientific and philosophical implications.

Our journey begins with our hero, who has now traversed countless parallel universes and experienced realities beyond human comprehension. He stands on the precipice of a new era for his kind - an era where humans are no longer confined to their home planet or even their home universe.

The first step towards becoming an intergalactic species involves mastering space travel within our own universe. We've already made significant strides in this area with advancements in rocket technology and plans for manned missions to Mars. However, these efforts pale compared to what would be required for inter-universal travel.

Consider Elon Musk's SpaceX program or NASA's Artemis project; they represent monumental steps forward but are still essentially baby steps when viewed against the backdrop of infinite universes. To leap from one universe to another requires not just advanced technology but also a radical shift in understanding physics and reality itself.

Let's take a moment here to ponder upon Albert Einstein's theory of relativity which revolutionized our understanding of space-time continuum. It was through his groundbreaking work that concepts like wormholes – shortcuts through spacetime – entered scientific parlance. Today, such ideas form the bedrock upon which theories about interdimensional travel rest.

But it isn't enough merely to reach other universes; we must also adapt physically and mentally once we get there—a challenge that brings us face-to-face with evolution itself.

Imagine landing on a planet where gravity is three times stronger than Earth's—our current physical form wouldn't cope well under such conditions over time without some level of adaptation or modification.

This leads us into controversial territory: genetic engineering and transhumanism. These fields suggest that we could use technology to enhance our physical and cognitive abilities, allowing us to survive in environments that would otherwise be inhospitable.

* * *

Our hero, for instance, has had his fair share of encounters with alien species who have adapted in ways unimaginable to humans. He's met beings made entirely of energy, entities capable of surviving without oxygen or sunlight, and creatures whose consciousness exists across multiple dimensions simultaneously.

But evolution isn't just about survival—it also involves the development of society and culture. As an intergalactic species, humanity would need a way to maintain unity amidst staggering diversity. This is where philosophy enters the picture.

The philosopher Immanuel Kant proposed a "universal law" principle —essentially stating that one should act only according to rules they'd wish everyone else followed as well. In an intergalactic society spanning countless worlds and cultures, such universal principles might serve as a unifying force.

In conclusion: evolving into an intergalic species requires advancements not just technologically but biologically and philosophically too.

As you ponder these ideas further here are some questions for reflection:

1) What kind of ethical issues might arise from genetic engineering aimed at adapting humans for life on other planets?

2) How can we ensure unity among diverse human colonies spread across different universes?

3) What role does Earth play in this grand vision? Is it still home or merely where we started?

And finally – imagine yourself standing alongside our hero on the brink of this new era; what do you see? How do you feel? And most importantly - are you ready?

This chapter serves as both a thought experiment and call-to-action because ultimately – the future belongs not just to those who dream but those willing enough to turn their dreams into reality!

INTEGRATING KNOWLEDGE FROM MULTIPLE DISCIPLINES

Chapter 53: Integrating Knowledge from Multiple Disciplines

In our journey through the multiverse, we've encountered countless realities and have been privy to an array of knowledge that transcends what we previously thought possible. Our hero has navigated through different dimensions, battled cross-dimensional crises, and even become a king amongst many worlds. But now comes perhaps one of the most challenging tasks yet - integrating this wealth of knowledge across multiple disciplines.

The exploration of parallel universes is not just a matter for physicists or cosmologists; it requires an interdisciplinary approach that draws on fields as diverse as philosophy, biology, computer science, psychology and even art. Each discipline offers unique insights into understanding the nature of reality and our place within it.

Consider how philosophy can help us grapple with the ethical dilemmas posed by existence in a multiverse. How should we act if every decision creates a new universe? What does morality look like when there are infinite versions of ourselves making different choices?

Biology could provide insights into how life might evolve under different physical laws or environmental conditions. Imagine discovering organisms that use dark matter instead of carbon to build their bodies or ecosystems based on principles entirely alien to us!

Computer science plays an integral role too – especially in terms of simulation theory which proposes that our universe could be nothing more than a sophisticated digital construct. This idea challenges our perception about consciousness and reality itself.

Psychology helps us understand how humans (and potentially other sentient beings) might perceive and interact with these alternate realities. It also aids in addressing existential questions raised by the concept of multiple selves existing simultaneously across various universes.

Artistic interpretations offer another valuable perspective – they allow us to visualize abstract concepts related to parallel universes in ways words often fail to capture adequately.

Our hero's journey so far has shown him firsthand how interconnected everything truly is - from quantum particles vibrating strings in eleven-dimensional space-time fabric to conscious entities pondering their existence across multiple realities. He has learned to appreciate the beauty of diversity in all its forms and understand that knowledge is not compartmentalized but rather a vast, interconnected web.

Now, he faces the task of integrating this newfound understanding into his worldview - a process that requires humility, open-mindedness, and an insatiable curiosity. It's about recognizing patterns between seemingly disparate fields and synthesizing them into a coherent whole.

As we near the end of our journey through the multiverse, let us take inspiration from our hero's quest for integration. Let us strive to break down barriers between disciplines and foster collaboration because it is only by doing so can we hope to unravel the mysteries of our boundless cosmos.

So here's your challenge: Think about how different areas of study could contribute to understanding parallel universes. How might they intersect? What unique insights could they offer? And most importantly – how can you apply this integrated approach in your own life?

Remember: The universe (or should I say multiverse?) is full of infinite possibilities waiting to be explored!

www.ingramcontent.com/pod-product-compliance
Lightning Source LLC
Chambersburg PA
CBHW062335290526
45794CB00005B/2045